Modern Birkhäuser Classics

Many of the original research and survey monographs, as well as textbooks, in pure and applied mathematics published by Birkhäuser in recent decades have been groundbreaking and have come to be regarded as foundational to the subject. Through the MBC Series, a select number of these modern classics, entirely uncorrected, are being re-released in paperback (and as eBooks) to ensure that these treasures remain accessible to new generations of students, scholars, and researchers.

The Implicit Function Theorem

History, Theory, and Applications

Steven G. Krantz
Harold R. Parks

Reprint of the 2003 Edition

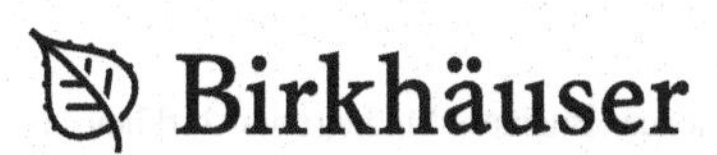

Steven G. Krantz
Department of Mathematics
Washington University
St. Louis, MO
USA

Harold R Parks
Department of Mathematics
Oregon State University
Corvallis, OR
USA

ISBN 978-1-4614-5980-4 ISBN 978-1-4614-5981-1 (eBook)
DOI 10.1007/978-1-4614-5981-1
Springer New York Heidelberg Dordrecht London

Library of Congress Control Number: 2012952321

Printed on acid-free paper

Springer is part of Springer Science+Business Media (www.birkhauser-science.com)

Steven G. Krantz
Harold R. Parks

The Implicit Function Theorem

History, Theory, and Applications

Birkhäuser
Boston • Basel • Berlin

Steven G. Krantz
Department of Mathematics
Washington University
St. Louis, MO 63130-4899
U.S.A.

Harold R. Parks
Department of Mathematics
Oregon State University
Corvallis, OR 97331-4605
U.S.A.

Library of Congress Cataloging-in-Publication Data

Krantz, Steven G. (Steven George), 1951-
The implicit function theorem : history, theory, and applications / Steven G. Krantz and Harold R. Parks.
p. cm.
Includes bibliographical references and index.
ISBN 0-8176-4285-4 (acid-free paper) – ISBN 3-7643-4285-4 (acid-free paper)
1. Implicit functions. I. Parks, Harold R., 1949- II. Title.

QA331.5.K7136 2002
515'.8–dc21

2002018234
CIP

AMS Subject Classifications: Primary: 26B10; Secondary: 01-01, 01-02, 53B99, 53C99, 35A99, 35B99

Printed on acid-free paper.

Birkhäuser 

ISBN 0-8176-4285-4 SPIN 10938065
ISBN 3-7643-4285-4

Reformatted from authors' files by TEXniques, Inc., Cambridge, MA.
Printed in the United States of America.

9 8 7 6 5 4 3 2

Birkhäuser Boston • Basel • Berlin
A member of BertelsmannSpringer Science+Business Media GmbH

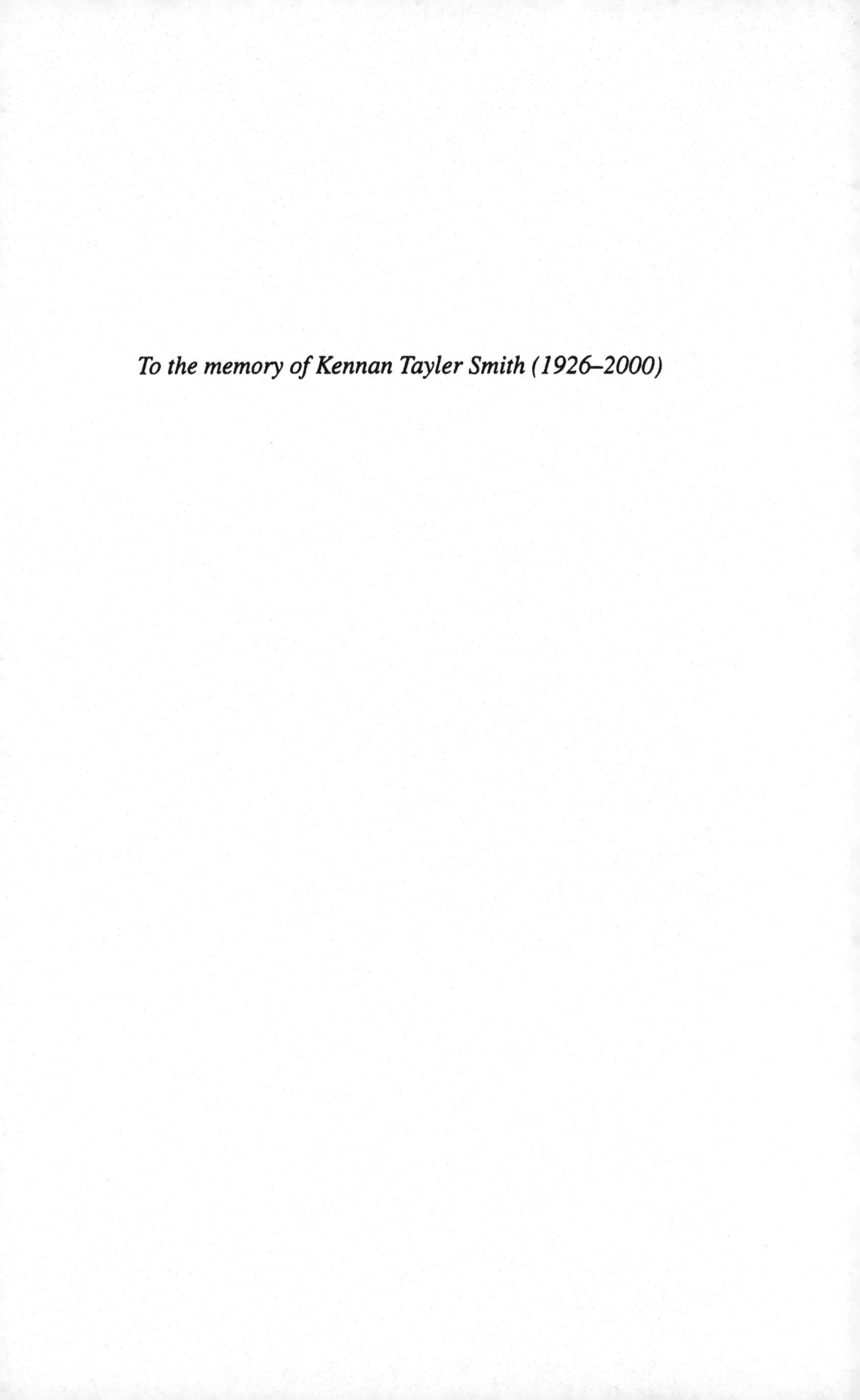

To the memory of Kennan Tayler Smith (1926–2000)

Contents

Preface

The implicit function theorem is, along with its close cousin the inverse function theorem, one of the most important, and one of the oldest, paradigms in modern mathematics. One can see the germ of the idea for the implicit function theorem in the writings of Isaac Newton (1642–1727), and Gottfried Leibniz's (1646–1716) work explicitly contains an instance of implicit differentiation. While Joseph Louis Lagrange (1736–1813) found a theorem that is essentially a version of the inverse function theorem, it was Augustin-Louis Cauchy (1789–1857) who approached the implicit function theorem with mathematical rigor and it is he who is generally acknowledged as the discoverer of the theorem. In Chapter 2, we will give details of the contributions of Newton, Lagrange, and Cauchy to the development of the implicit function theorem.

The form of the implicit function theorem has evolved. The theorem first was formulated in terms of complex analysis and complex power series. As interest in, and understanding of, real analysis grew, the real-variable form of the theorem emerged. First the implicit function theorem was formulated for functions of two real variables, and the hypothesis corresponding to the Jacobian matrix being nonsingular was simply that one partial derivative was nonvanishing. Finally, Ulisse Dini (1845–1918) generalized the real-variable version of the implicit function theorem to the context of functions of any number of real variables. As mathematicians understood the theorem better, alternative proofs emerged, and the associated modern techniques have allowed a wealth of generalizations of the implicit function theorem to be developed.

Today we understand the implicit function theorem to be an *ansatz*, or a way of looking at problems. There are implicit function theorems, inverse function theorems, rank theorems, and many other variants. These theorems are valid on

Euclidean spaces, manifolds, Banach spaces, and even more general settings. Roughly speaking, the implicit function theorem is a device for solving equations, and these equations can live in many different settings.

In addition, the theorem is valid in many categories. The textbook formulation of the implicit function theorem is for C^k functions. But in fact the result is true for $C^{k,\alpha}$ functions, Lipschitz functions, real analytic functions, holomorphic functions, functions in Gevrey classes, and for many other classes as well. The literature is rather opaque when it comes to these important variants, and a part of the present work will be to set the record straight.

Certainly one of the most powerful forms of the implicit function theorem is that which is attributed to John Nash (1928–) and Jürgen Moser (1928–1999). This device is actually an infinite iteration scheme of implicit function theorems. It was first used by John Nash to prove his celebrated imbedding theorem for Riemannian manifolds. Jürgen Moser isolated the technique and turned it into a powerful tool that is now part of partial differential equations, functional analysis, several complex variables, and many other fields as well. This text will culminate with a version of the Nash–Moser theorem, complete with proof.

This book is one both of theory and practice. We intend to present a great many variants of the implicit function theorem, complete with proofs. Even the important implicit function theorem for real analytic functions is rather difficult to pry out of the literature. We intend this book to be a convenient reference for all such questions, but we also intend to provide a compendium of examples and of techniques. There are applications to algebra, differential geometry, manifold theory, differential topology, functional analysis, fixed point theory, partial differential equations, and to many other branches of mathematics. One learns mathematics (in part) by watching others do it. We hope to set a suitable example for those wishing to learn the implicit function theorem.

The book should be of interest to advanced undergraduates, graduate students, and professional mathematicians. Prerequisites are few. It is not necessary that the reader be already acquainted with the implicit function theorem. Indeed, the first chapter provides motivation and examples that should make clear the form and function of the implicit function theorem. A bit of knowledge of multivariable calculus will allow the reader to tackle the elementary proofs of the implicit function theorem given in Chapter 3. Rudiments of real and functional analysis are needed for the third proof in Chapter 3 which uses the Contraction Mapping Fixed Point Principle. Some knowledge of complex analysis is required for a complete reading of the historical material—this seems to be unavoidable since the earliest rigorous work on the implicit function theorem was formulated in the context of complex variables. In many cases a willing suspension of disbelief and a bit of determination will serve as a thorough grounding in the basics.

There are many sophisticated applications of implicit function theorems, particularly the Nash–Moser theorem, in modern mathematics. The imbedding theorem for Riemannian manifolds, the imbedding theorem for CR manifolds, and the deformation theory of complex structures are just a few of them. Richard Hamilton's masterful survey paper (see the Bibliography) indicates several more applications from different parts of mathematics. While each of these is a lovely *tour de force* of modern analytical technique, it is also the case that each requires considerable technical background. In order to keep the present volume as self-contained as possible, we have decided not to include any of these modern applications; instead we have provided exclusively classical applications of the implicit function theorem. For a basic book on the subject, we have found this choice to be most propitious.

We intend this book to be a useful resource for scientists of all types. We have exerted a considerable effort to make the bibliography extensive (if not complete). Therefore topics that can only be touched on here can be amplified with further reading. Although there are no formal exercises, the extensive remarks provide grist for further thought and calculation. We trust that our exposition will imbue our readers with some of the same fascination that led to the writing of this book.

There are a number of people whom we are pleased to thank for their helpful comments and contributions: David Barrett, Michael Crandall, John P. D'Angelo, Gerald B. Folland, Judith Grabiner, Robert E. Greene, Lars Hörmander, Seth Howell, Kang-Tae Kim, Laszlo Lempert, Maurizio Letizia, Richard Rochberg, Walter Rudin, Steven Weintraub, Dean Wills, Hung-Hsi Wu. Robert Burckel cast his critical eye on every page of our manuscript and the result is a much cleaner and more accurate book. Librarian Barbara Luszczynska performed yeoman service in helping us to track down references. This book is better because of the friendly assistance of all these good people; but, of course, all remaining failings are the province of the authors.

Washington University, St. Louis — *Steven G. Krantz*
Oregon State University, Corvallis — *Harold R. Parks*

1

Introduction to the Implicit Function Theorem

1.1 Implicit Functions

To the beginning student of calculus, a function is given by an analytic expression such as

$$f(x) = x^3 + 2x^2 - x - 3\,, \tag{1.1}$$

$$g(y) = \sqrt{y^2 + 1}\,, \tag{1.2}$$

or

$$h(t) = \cos(2\pi t)\,. \tag{1.3}$$

In fact, 250 years ago this was the approach taken by Léonard Euler (1707–1783) when he wrote (see Euler [EB 88]):

> *A function of a variable quantity is an analytic expression composed in any way whatsoever of the variable quantity and numbers or constant quantities.*

Almost immediately, one finds the notion of "function as given by a formula" to be too limited for the purposes of calculus. For example, the locus of

$$y^5 + 16y - 32x^3 + 32x = 0 \tag{1.4}$$

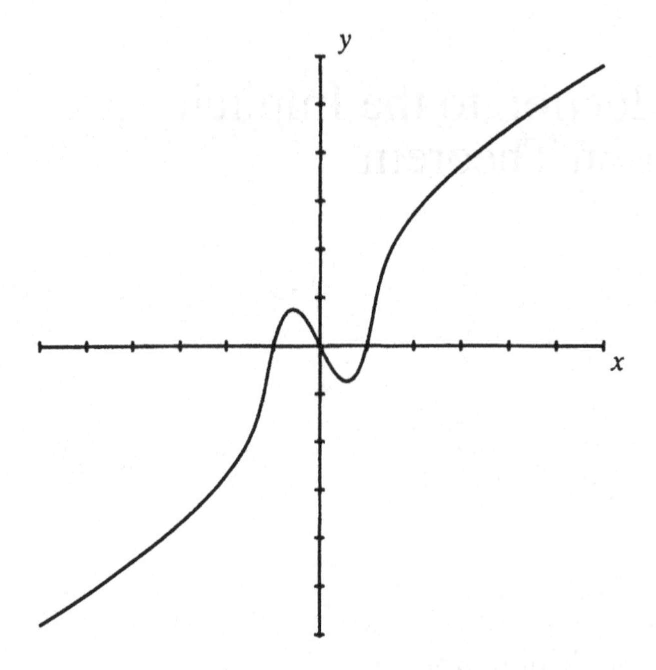

Figure 1.1. The Locus of Points Satisfying (1.4)

defines the nice subset of $\mathbf{R}^2$ that is sketched in Figure 1.1. The figure leads us to suspect that the locus is the graph of y as a function of x, but no formula for that function exists.

In contrast to the naive definition of functions as formulas, the modern, set-theoretic definition of a function is formulated in terms of the graph of the function. Precisely, a *function* with *domain* X and *codomain* or *range* Y is a subset, let us call it f, of the cartesian product

$$X \times Y = \{(x, y) : x \in X,\ y \in Y\}$$

having the properties that (i) for each $x \in X$ there is an element $(x, y) \in f$, and (ii) if $(x, y) \in f$ and $(x, \tilde{y}) \in f$, then $y = \tilde{y}$. In case these two properties hold, the choice of $x \in X$ determines the unique $y \in Y$ for which $(x, y) \in f$; because of this uniqueness, we find it a convenient shorthand to write

$$y = f(x)$$

to mean that $(x, y) \in f$.

Example 1.1.1 The locus defined by (1.4) has the property that, for each choice of $x \in \mathbf{R}$, there is a unique $y \in \mathbf{R}$ such that the pair (x, y) satisfies the equation. Thus there is a function, f, in the modern sense, such that the graph $y = f(x)$ is the locus of (1.4).

To confirm this assertion, we fix a value of $x \in \mathbf{R}$ and consider the left-hand side of (1.4) as a function of y alone. That is, we will examine the behavior of

$$F(y) = y^5 + 16y - 32x^3 + 32x$$

with x fixed.

Since the powers of y in $F(y)$ are odd, we have $\lim_{y\to-\infty} F(y) = -\infty$ and $\lim_{y\to+\infty} F(y) = +\infty$. Also we have

$$F'(y) = 5y^4 + 16 > 0,$$

so $F(y)$ is strictly increasing as y increases. By the intermediate value theorem, we see that $F(y)$ attains the value 0 for a unique value of y. That value of y is the value of $f(x)$ for the fixed value of x under consideration. □

Note that it is not clear from (1.4) by itself that y is a function of x. Only by doing the extra work in the example can we be certain that y really is uniquely defined as a function of x. Because it is not immediately clear from the defining equation that a function has been given, we say that the function is defined *implicitly* by (1.4). In contrast, when we see

$$y = f(x) \tag{1.5}$$

written, we then take it as a hypothesis that $f(x)$ is a function of x; no additional verification is required, even when in the right-hand side the function is simply a symbolic representation as in (1.5) rather than a formula as in (1.1), (1.2), and (1.3). To distinguish them from implicitly defined functions, the functions in (1.1), (1.2), (1.3), and (1.5) are called (in this book) *explicit* functions.

1.2 An Informal Version of the Implicit Function Theorem

Thinking heuristically, one usually expects that one equation in one variable

$$F(x) = c,$$

c a constant, will be sufficient to determine the value of x (though the existence of more than one, but only finitely many, solutions would come as no surprise).[1] When there are two variables, one expects that it will take two simultaneous equations

$$\begin{aligned} F(x, y) &= c, \\ G(x, y) &= d, \end{aligned}$$

[1] What we are doing is informally describing the notion of "degrees of freedom" that is commonly used in physics.

c and d constants, to determine the values of both x and y. In general, one expects that a system of m equations in m variables

$$\begin{aligned} F_1(x_1, x_2, \ldots, x_m) &= c_1, \\ F_2(x_1, x_2, \ldots, x_m) &= c_2, \\ &\vdots \\ F_m(x_1, x_2, \ldots, x_m) &= c_m, \end{aligned} \tag{1.6}$$

$c_1, c_2, \ldots, c_m$ constants, will be just the right number of equations to determine the values of the variables. But of course we must beware of redundancies among the equations. That is, we must check that the system is nondegenerate—in the sense that a certain determinant does not vanish.

In case the equations in (1.6) are all linear, we can appeal to linear algebra to make our heuristic thinking precise (see any linear algebra textbook): A necessary and sufficient condition to guarantee that (1.6) has a unique solution for all values of the constants c_i is that the matrix of coefficients of the linear system has rank m.

We continue to think heuristically: If there are more variables than equations in our system of simultaneous equations, say

$$\begin{aligned} F_1(x_1, x_2, \ldots, x_n) &= c_1, \\ F_2(x_1, x_2, \ldots, x_n) &= c_2, \\ &\vdots \\ F_m(x_1, x_2, \ldots, x_n) &= c_m, \end{aligned} \tag{1.7}$$

where the c's are still constants and where $n > m$, then we would hope to treat those $n-m$ extra variables as parameters, thereby forcing m of the variables to be implicit functions of the $n - m$ parameters. Again, in the case of linear functions, the situation is well understood: As long as the matrix of coefficients has rank m, it will be possible to express some set of m of the variables as functions of the other $n - m$ variables. Moreover, for any set of m independent columns of the matrix of coefficients of the linear system, the corresponding m variables can be expressed as functions of the other variables.

In the general case, as opposed to the linear case, the system of equations (1.7) defines a completely arbitrary subset of $\mathbf{R}^n$ (an arbitrary closed subset if the functions are continuous). Only under special conditions will (1.7) define m of the variables to be implicit functions of the other $n - m$ variables. It is the purpose of the implicit function theorem to provide us with a powerful method, or collection of methods, for insuring that we are in one of those special situations for which the heuristic argument is correct.

The implicit function theorem is grounded in differential calculus; and the bedrock of differential calculus is linear approximation. Accordingly, one works in a neighborhood of a point $(p_1, p_2, \ldots, p_n)$, where the equations in (1.7) all hold at $(p_1, p_2, \ldots, p_n)$ and where the functions in (1.7) can all be linearly approximated by their differentials. We are now in a position to state the implicit function theorem in informal terms (we shall give a more formal enunciation later):

> **(Informal) Implicit Function Theorem** *Let the functions in (1.7) be continuously differentiable. If (1.7) holds at $(p_1, p_2, \ldots, p_n)$ and if, when the functions in (1.7) are replaced by their linear approximations, a particular set of m variables can be expressed as functions of the other $n-m$ variables, then, for (1.7) itself, the same m variables can be defined to be implicit functions of the other $n-m$ variables in a neighborhood of $(p_1, p_2, \ldots, p_n)$. Additionally, the resulting implicit functions are continuously differentiable and their derivatives can be computed by implicit differentiation using the familiar method learned as part of the calculus.*

Let us look at a very simple example in which there is only one, well-understood, equation in two variables. We will treat this example in detail for the benefit of the reader who is not already comfortable with the ideas we have been discussing.

Example 1.2.1 Consider

$$x^2 + y^2 = 1\,. \tag{1.8}$$

The locus of points defined by (1.8) is the circle of radius 1 centered at the origin. Of course, in a suitable neighborhood of any point $P = (p, q)$ satisfying (1.8) and for which $q \neq 0$, we can solve the equation to express y explicitly as

$$y = \pm\sqrt{1 - x^2}\,,$$

where the choice of $+$ or $-$ is dictated by whether q is positive or negative. (Likewise, we could just as easily have dealt with the case in which $p \neq 0$ by solving for x as an explicit function of y.)

The usefulness of the implicit function theorem stems from the fact that we can avoid explicitly solving the equation. To take the point of view of the implicit function theorem, we linearly approximate the left-hand side of (1.8). In a neighborhood of a point $P = (p, q)$, a continuously differentiable function $F(x, y)$ is linearly approximated by

$$a\,\Delta x + b\,\Delta y + c\,,$$

where a is the value of $\partial F/\partial x$ evaluated at P, Δx is the change in x made in going from $P = (p, q)$ to the point (x, y), b is the value of $\partial F/\partial y$ evaluated at P, Δy is the change in y made in going from $P = (p, q)$ to the point (x, y), and c is the value of F at P. In this example, $F(x, y) = x^2 + y^2$, the left-hand side of (1.8).

We compute

$$\frac{\partial}{\partial x}\left(x^2 + y^2\right)\Big|_{(x,y)=(p,q)} = 2p$$

and

$$\frac{\partial}{\partial y}\left(x^2 + y^2\right)\bigg|_{(x,y)=(p,q)} = 2q\,.$$

Thus, in a neighborhood of the point $P = (p, q)$ which satisfies (1.8), the left-hand side of (1.8) is linearly approximated by

$$(2p)\,(x - p) + (2q)\,(y - q) + 1 = 2px + 2qy - 1\,.$$

When we replace the left-hand side of (1.8) by its linear approximation and simplify we obtain

$$px + qy = 1\,, \tag{1.9}$$

which, of course, is the equation of the tangent line to the circle at the point P.

The implicit function theorem tells us that whenever we can solve the approximating linear equation (1.9) for y as a function of x, then the original equation (1.8) defines y implicitly as a function of x. Clearly, we can solve (1.9) for y as a function of x exactly when $q \neq 0$, so it is in this case that the implicit function theorem guarantees that (1.8) defines y as an implicit function of x. This agrees perfectly with what we found when we solved the equation explicitly. □

Remark 1.2.2 Looking at the circle, we see that it is impossible to use (1.8) to define y as a function of x in any open interval around $x = 1$ or in any open interval around $x = -1$. For other equations, an implicit function may happen to exist in a neighborhood of a point at which the implicit function theorem does not apply but, in such a case, the function may or may not be differentiable.

An example in which there are three variables and two equations will serve to illustrate the connection between linear algebra and the implicit function theorem.

Example 1.2.3 Fix $R \geq \sqrt{2}$ and consider the pair of equations

$$\begin{aligned} x^2 + y^2 + z^2 &= R^2\,, \\ xy &= 1 \end{aligned} \tag{1.10}$$

near the point $P = (1, 1, \rho)$, where $\rho = \sqrt{R^2 - 2}$.

We could solve the system explicitly. But it is instructive to instead take the point of view of the implicit function theorem. There are three variables and two equations, so the heuristic argument above tells us to expect two variables to be implicit functions of the third.

Computing partial derivatives and evaluating at $(1, 1, \rho)$ to linearly approximate the functions in (1.10), we obtain the equations

$$\begin{aligned} x + y + \rho z &= 2 + \rho^2\,, \\ x + y &= 2\,. \end{aligned} \tag{1.11}$$

This system of equations is the linearization of the original system. The first equation in (1.11) defines the tangent plane at P of the locus defined by the first equation in (1.10) and the second equation in (1.11) defines the tangent plane at the same point of the locus defined by the second equation in (1.10). Clearly, the two tangent planes have a non-trivial intersection because both automatically contain the point P.

The requirement that needs to be verified before the implicit function theorem can be applied is that we can solve the linear system (1.11) for two of the variables as a function of the third. Geometrically, this corresponds to showing that the intersection of the tangent planes is a line, because it is along a line in $\mathbf{R}^3$ that two of the variables can be expressed as a function of the third.

We now appeal to linear algebra. The matrix of coefficients for the linear system is

$$D = \begin{pmatrix} 1 & 1 & \rho \\ 1 & 1 & 0 \end{pmatrix} .$$

The necessary and sufficient condition for being able to solve (1.11) for two of the variables as a function of the third is that D have rank 2. Clearly, the rank of D is 2 if and only if $\rho \neq 0$. Thus, when $R > \sqrt{2}$, the implicit function theorem then guarantees that some pair of the variables can be defined implicitly in terms of the remaining variable.

On the other hand, when $\rho = 0$, or equivalently when $R = \sqrt{2}$, the rank of D is 1 and the implicit function theorem does *not* apply. Not only does the implicit function theorem not apply, but it is easy to see that $(1, 1, 0)$ and $(-1, -1, 0)$ are the only solutions of (1.10).

Assume now that $\rho \neq 0$. The implicit function theorem tells us that if we can solve the linear system (1.11) for a particular pair of the variables in terms of the third, then the original system of equations defines the same two variables as implicit functions of the third near $(1, 1, \rho)$. To determine which pairs of variables are functions of the third, we again appeal to linear algebra. Any two independent columns of D will correspond to variables in (1.11) that can be expressed as functions of the third. Thus, the implicit function theorem gives us the pair $x(y)$ and $z(y)$ satisfying (1.10), or the pair $y(x)$ and $z(x)$ satisfying (1.10).

In this example, not only does the implicit function theorem *not* allow us to assert the existence of $x(z)$ and $y(z)$ satisfying (1.10), but no such functions exist. □

1.3 The Implicit Function Theorem Paradigm

In the last section, we described the heuristic thinking behind the implicit function theorem and stated the theorem in informal terms. Even though the heuristic argument behind the result is rather simple, the implicit function theorem is a fundamental and powerful part of the foundation of modern mathematics. Originally conceived over two hundred years ago as a tool for studying celestial mechanics (see also Section 2.3), the implicit function theorem now has many formulations and is used in many parts of mathematics. Virtually every category of functions has its own special version of the implicit function theorem, and there are particular versions adapted to Banach spaces, algebraic geometry, various types of geometrically degenerate situations, and to functions that are not even smooth. Some of these are quite sophisticated, and have been used in startling ways to solve important open problems (the imbedding problem for Riemannian manifolds and the imbedding problem for CR manifolds are just two of them).

The implicit function theorem paradigm: *Given three topological spaces $\mathbb{X}$, $\mathbb{Y}$, $\mathbb{Z}$ (these spaces need not be distinct), a continuous function $\mathcal{F} : \mathbb{X} \times \mathbb{Y} \to \mathbb{Z}$, and points $X_0 \in \mathbb{X}$, $Y_0 \in \mathbb{Y}$, $Z_0 \in \mathbb{Z}$ such that*

$$\mathcal{F}(X_0, Y_0) = Z_0 \,,$$

an implicit function theorem must describe an appropriate nondegeneracy condition on $\mathcal{F}$ at (X_0, Y_0) sufficient to imply the existence of neighborhoods U of X_0 in $\mathbb{X}$, V of Y_0 in $\mathbb{Y}$, and of a function $F : U \to V$ satisfying the following two conditions:

$$\begin{aligned} F(X_0) &= Y_0 \,, \\ \mathcal{F}[X, F(X)] &= Z_0, \text{ for all } X \in U \,. \end{aligned} \tag{1.12}$$

Additionally, an implicit function theorem will entail the conclusion that the function F is well behaved *in some appropriate sense, and it is usually an important part of the theorem that F is the unique function satisfying (1.12).*

The simplest case of the above paradigm is to let all three of the topological spaces be the real numbers $\mathbf{R}$. The function $\mathcal{F}$ is assumed to be continuously differentiable and the nondegeneracy condition is the nonvanishing of the partial derivative with respect to Y. We now state the result formally as a theorem.

Theorem 1.3.1 *Let $\mathcal{F}$ be a real-valued continuously differentiable function defined in a neighborhood of $(X_0, Y_0) \in \mathbf{R}^2$. Suppose that $\mathcal{F}$ satisfies the two conditions*

$$\mathcal{F}(X_0, Y_0) = Z_0 \,,$$
$$\frac{\partial \mathcal{F}}{\partial Y}(X_0, Y_0) \neq 0 \,.$$

Then there exist open intervals U and V, with $X_0 \in U$, $Y_0 \in V$, and a unique function $F : U \to V$ satisfying

$$\mathcal{F}[X, F(X)] = Z_0, \text{ for all } X \in U \,,$$

and this function F is continuously differentiable with

$$\frac{dY}{dX}(Y_0) = F'(Y_0) = -\left[\frac{\partial \mathcal{F}}{\partial X}(X_0, Y_0)\right] \bigg/ \left[\frac{\partial \mathcal{F}}{\partial Y}(X_0, Y_0)\right] \,. \tag{1.13}$$

Because this theorem involves partial derivatives, the theorem *per se* is not usually taught in a first calculus course. Instead, a disguised form of Equation (1.13) is taught: The student is told to go ahead and differentiate $\mathcal{F}(X, Y) = Z_0$ with respect to X using the chain rule and assuming that dY/dX exists. If it is then possible to solve for dY/dX when $X = X_0$ and $Y = Y_0$, the student is assured that the result is correct (as the theorem in fact guarantees). This somewhat *ad hoc* process is called *implicit differentiation*. Once the beginning student of calculus has learned about partial differentiation, Theorem 1.3.1 is likely to be the first version of the implicit function theorem presented.

By approaching this basic freshman calculus version of the implicit function theorem via the paradigm, we see that a natural generalization would arise by replacing **R** by **C** (that generalization is stated and proved in Section 2.4). In fact, there is no limit to the number of variations that can be made on this theme by altering the choice of topological spaces, or the category of functions considered, or the type of nondegeneracy conditions used, or the conclusions about what is a "well behaved" implicit function.

A corollary of Theorem 1.3.1 is obtained by setting

$$\mathcal{F}(X, Y) = X - G(Y),$$

with $G : \mathbf{R} \to \mathbf{R}$ a continuously differentiable function. The nondegeneracy requirement becomes $G'(Y_0) \neq 0$. Taking $Z_0 = 0$ and assuming $X_0 = G(Y_0)$, Theorem 1.3.1 guarantees the existence of a function F satisfying

$$G[F(X)] = X,$$

that is, F is the inverse function to G. We also conclude that

$$F'(X_0) = 1\big/G'(Y_0).$$

This result is the inverse function theorem taught in freshman calculus.

Both the implicit function theorem and the inverse function theorem might be proved in an honors course in calculus, but most students will first see the proofs in a course on advanced calculus. Nonetheless, a student will probably never really apply the theorems until more advanced mathematical work.

Example 1.3.2 Consider the equation

$$x = y - \epsilon \sin(y), \tag{1.14}$$

where ϵ is a small constant. While the notation we are using is different, (1.14) has the same form as Kepler's equation in celestial mechanics. A classical problem was to solve (1.14) for y as a function of x, that is, to find the inverse function. This cannot be done in closed form using elementary functions, but a positive result can be obtained using infinite series. The resulting formula is known as the Lagrange inversion theorem. All of this is discussed in more detail in Section 2.3. Here we note that

$$\frac{d}{dy}\Big[y - \epsilon \sin(y)\Big] = 1 - \epsilon \cos(y) \neq 0$$

holds, provided $|\epsilon| < 1$. Thus, the simple freshman calculus form of the inverse function theorem described above applies. □

In general, the implicit function theorem and the inverse function theorem can be thought of as equivalent, companion formulations of the same basic idea. In any particular context, one may find it easier to take one approach or the other.

To continue our more formal presentation of the the implicit function theorem, we give a simple, if typical, formulation of the theorem. For convenience in this rather elementary introduction, we state the result in $\mathbf{R}^3$. Be assured that the implicit function theorem is true in any dimensional space—even in infinite dimensional spaces.

Theorem 1.3.3 *We let $U \subseteq \mathbf{R}^3$ be an open set and we assume that*

$$\mathcal{F} = (\mathcal{F}_1, \mathcal{F}_2) : U \to \mathbf{R}^2$$

is a continuously differentiable function. Further assume that, at a point $\mathbf{a} = (a_1, a_2, a_3) \in U$, it holds that $\mathcal{F}(\mathbf{a}) = 0$ and

$$\det \begin{pmatrix} \dfrac{\partial \mathcal{F}_1}{\partial x_2} & \dfrac{\partial \mathcal{F}_1}{\partial x_3} \\ \dfrac{\partial \mathcal{F}_2}{\partial x_2} & \dfrac{\partial \mathcal{F}_2}{\partial x_3} \end{pmatrix} \neq 0 .$$

Then there is a product neighborhood $V \times W \subseteq U$, with $a_1 \in V \subseteq \mathbf{R}$ and $(a_2, a_3) \in W \subseteq \mathbf{R}^2$, and a unique, continuously differentiable mapping

$$F = (F_1, F_2) : V \to W$$

such that $(a_2, a_3) = F(a_1)$ and, for each $x_1 \in V$, it holds that

$$\mathcal{F}[x_1, F_1(x_1), F_2(x_1)] = 0 .$$

Again, we will not prove this result here, but refer the reader to Section 3.3. This theorem applies to Example 1.2.3 of the preceding section.

In words, we can think of Theorem 1.3.3 in this way: Imagine a pair of equations in the variables x_1, x_2, x_3 that has the form

$$\begin{aligned} \mathcal{F}_1(x_1, x_2, x_3) &= 0 , \\ \mathcal{F}_2(x_1, x_2, x_3) &= 0 . \end{aligned}$$

We wish to solve for x_2 and x_3 in terms of the remaining variable x_1. Ideally, x_2 and x_3 should be expressed as smooth functions of x_1. The condition that will guarantee this conclusion is that the "derivative" with respect to the variables for which we wish to solve should be invertible. Here the "derivative" is a linear map from $\mathbf{R}^2$ to $\mathbf{R}^2$, so it is invertible if and only if the determinant is nonvanishing.

The next example of the implicit function theorem will lead to a corollary form of the inverse function theorem. In comparison with Theorem 1.3.3, all we really change is the dimension of the domain of $\mathcal{F}$.

Theorem 1.3.4 *We let $U \subseteq \mathbf{R}^4$ be an open set and we assume that*

$$\mathcal{F} = (\mathcal{F}_1, \mathcal{F}_2) : U \to \mathbf{R}^2$$

is a continuously differentiable function. Further assume that, at a point $\mathbf{a} = (a_1, a_2, a_3, a_4) \in U$, it holds that $\mathcal{F}(\mathbf{a}) = \mathbf{0}$ and

$$\det \begin{pmatrix} \dfrac{\partial \mathcal{F}_1}{\partial x_3} & \dfrac{\partial \mathcal{F}_1}{\partial x_4} \\ \dfrac{\partial \mathcal{F}_2}{\partial x_3} & \dfrac{\partial \mathcal{F}_2}{\partial x_4} \end{pmatrix} \neq 0 . \tag{1.15}$$

Then there is a product neighborhood $V \times W \subseteq U$, with $(a_1, a_2) \in V \subseteq \mathbf{R}^2$ and $(a_3, a_4) \in W \subseteq \mathbf{R}^2$, and a unique, continuously differentiable mapping

$$F = (F_1, F_2) : V \to W$$

such that $(a_3, a_4) = F(a_1, a_2)$ and, for each $\mathbf{x} = (x_1, x_2) \in V$, it holds that

$$\mathcal{F}[x_1, x_2, F_1(\mathbf{x}), F_2(\mathbf{x})] = \mathbf{0} .$$

Once more the result is a special case of those in Section 3.3.

Corollary 1.3.5 *We let $Y \subseteq \mathbf{R}^2$ be an open set and we assume that*

$$G = (G_1, G_2) : Y \to \mathbf{R}^2$$

is a continuously differentiable function. We further assume that, at a point $\mathbf{b} = (b_1, b_2) \in Y$, it holds that

$$\det \begin{pmatrix} \dfrac{\partial G_1}{\partial y_1} & \dfrac{\partial G_1}{\partial y_2} \\ \dfrac{\partial G_2}{\partial y_1} & \dfrac{\partial G_2}{\partial y_2} \end{pmatrix} \neq 0 . \tag{1.16}$$

Then there are neighborhoods $V, W \subseteq \mathbf{R}^2$, with $\mathbf{a} = (a_1, a_2) = G(\mathbf{b}) \in V$ and $\mathbf{b} \in W$ and a unique, continuously differentiable mapping

$$F = (F_1, F_2) : V \to W$$

such that $\mathbf{b} = F(\mathbf{a})$ and, for each $\mathbf{x} = (x_1, x_2) \in V$, it holds that

$$\mathbf{x} = G[F(\mathbf{x})] .$$

Proof. We define $\mathcal{F} : \mathbf{R}^2 \times Y \to \mathbf{R}^2$ by setting

$$\mathcal{F}(x_1, x_2, x_3, x_4) = (x_1, x_2) - G(x_3, x_4) .$$

Equation (1.16) implies that (1.15) holds at (a_1, a_2, b_1, b_2). Thus the corollary follows from Theorem 1.3.4. □

In the next example, we show how the implicit function theorem, in the form of Corollary 1.3.5 can be applied to the study of a partial differential equation.

Example 1.3.6 Let W be an open set in $\mathbf{R}^2$ and let $u : W \to \mathbf{R}$ be a twice continuously differentiable function. If at a point $(x_0, y_0) \in W$ we know that

$$u_{xx}u_{yy} - u_{xy}^2 \neq 0 \tag{1.17}$$

holds, where the subscripts indicate partial differentiation, then, in a neighborhood of $(x_0, y_0) \in W$, one can make an invertible transformation from (x, y) to (ξ, η) and define a function $\omega(\xi, \eta)$ so that the formulas

$$\begin{aligned} \omega(\xi, \eta) + u(x, y) &= x\xi + y\eta, \\ \xi = u_x, &\quad \eta = u_y, \\ x = \omega_\xi, &\quad y = \omega_\eta \end{aligned} \tag{1.18}$$

hold.

To see that such a transformation can be made, we apply Corollary 1.3.5 to the function from $\mathbf{R}^2$ to $\mathbf{R}^2$ given by

$$(x, y) \mapsto \Big(u_x(x, y), u_y(x, y)\Big).$$

Equation (1.17) is exactly the hypothesis needed to apply Corollary 1.3.5 to conclude that the transformation

$$\begin{aligned} \xi &= u_x(x, y) \\ \eta &= u_y(x, y) \end{aligned}$$

is invertible.

Defining ω by setting

$$\omega(\xi, \eta) = -u(x, y) + x\xi + y\eta,$$

we compute

$$\begin{aligned} \omega_\xi &= -u_x x_\xi - u_y y_\xi + x_\xi \xi + x + y_\xi \eta \\ &= -\xi x_\xi - \eta y_\xi + x_\xi \xi + x + y_\xi \eta = x, \\ \omega_\eta &= -u_x x_\eta - u_y y_\eta + x_\eta \xi + y_\eta \eta + y \\ &= -\xi x_\eta - \eta y_\eta + x_\eta \xi + y_\eta \eta + y = y, \end{aligned}$$

showing that all the formulas in (1.18) hold. □

Remark 1.3.7 The transformation effected in the example is known as a *Legendre transformation* in honor of Adrien Marie Legendre (1752–1833) who introduced the idea in 1789. Such a transformation can sometimes be used to simplify the integration of a partial differential equation. Of course, Legendre transformations can be performed when there are more than two variables (see Courant–Hilbert [CH 62]). There are also sophisticated uses of Legendre transformations in mechanics (see Arnol'd [Ar 78]).

2

History

2.1 Historical Introduction

The earliest works on algebra beginning with *Al-jabr w'al muqâbala* by Mohammed ben Musa Al-Khowârizmî (circa A.D. 825), from whence we get the word "algebra" (and the word "algorithm"), presented problems and solutions by numerical example. The notion of a "function," whether explicit or implicit, would make no sense in such a context. It was not until about 1600 that the idea of using letters to denote both unknowns and coefficients was introduced by François Viète (1540–1603). The algebraic methods of Viète were taken up by René Descartes (1596–1650) and combined with Descartes's own coordinate system inspiration. That fundamental advance in 1637 finally brought mathematics to the point that the notion of a function could make sense. From the beginning, many of the functions were defined implicitly, as in the general quadratic curve

$$Ax^2 + Bxy + Cy^2 + Dx + Ey + F = 0$$

which arose in Descartes' solution of the Problem of Pappus (circa 300 A.D.), a problem that had been unsolved for more than a millennium.[1]

It appears that, before 1800, no one felt that there was a need to prove the existence of any implicit function. In fact, we can get a sense of their outlook from the words of Euler (as translated by John D. Blandon; see Euler [EB 88; page 5]):

[1] For more detail on these matters, see Hairer and Wanner [HW 96].

Indeed frequently algebraic functions cannot be expressed explicitly. For example, consider the function Z of z defined by the equation, $Z^5 = az^2Z^3 - bz^4Z^2 + cz^3Z - 1$. *Even if this equation cannot be solved, still it remains true that Z is equal to some expression composed of the variable z and constants, and for this reason Z shall be a function of z.*

The approach to implicit functions was to show how they behave, rather than to prove they exist. The work of Isaac Newton that we describe below may be one of the first instances of analyzing the behavior of an implicitly defined function. In the context of calculus, Gottfried Leibniz (1646–1716) applied implicit differentiation as early as 1684 (see [St 69; pages 276–278]).

In 1770, Joseph Lagrange proved what may be the first true implicit function theorem, but in its closely related form as an inverse function theorem. The result is now known as the Lagrange Inversion Theorem. Lagrange's theorem is what we would consider a special case of the inverse function theorem for formal power series.

Lagrange's theorem is quite important for celestial mechanics. Celestial mechanics occupied a central role in 18th and 19th century mathematics and Lagrange's theorem was very well known. Cauchy, in his quest to make mathematics rigorous, naturally gave his attention to that theorem and its generalizations. So it is that William Fogg Osgood (1864–1943), one of the first great American analysts,[2] attributes the implicit function theorem to Cauchy; more specifically, Osgood cites the "Turin Memoir" of Cauchy as the source of the implicit function theorem. The story of Cauchy's exile to Turin is a subject of some controversy and we will leave it to the reader to consult other sources, such as Belhoste [Be 91] and the references therein. In fact, there are two Turin Memoirs by Cauchy, and it is the first that contains the implicit function theorem. Also, we should note that the first Turin Memoir was, so to speak, printed, but not published; that is, while all parts of the first Turin Memoir ultimately appear in Cauchy's collected works, the memoir as a unified whole does not; nonetheless, the portion of the first Turin Memoir containing the implicit function theorem can be found in Cauchy [Ca 16].

It was only later in the 19th century that the profound differences between complex analysis and real analysis came to be more fully appreciated. Thus the real-variable form of the implicit function theorem was not enunciated and proved until the work of Ulisse Dini (1845–1918) that was first presented at the University of Pisa in the academic year 1876–1877 (see Dini [Di 07]).

In the remainder of this chapter, we will describe the contributions of Newton, Lagrange, and Cauchy mentioned above. The real-variable approach, going back to Dini, is pervasive throughout the rest of this book.

---

[2] Admittedly, he did earn his Ph.D. in Europe (at Heidelberg under Max Noether (1844–1921)).

2.2 Newton

The basic problem addressed by the implicit function theorem is of such fundamental interest that the genesis of the theorem goes back to Newton. In the Latin manuscript *De Analysi per Æquationes Infinitas* of 1669[3], Newton addresses the question of expressing the solution of the equation

$$Y^3 + a^2 Y - 2a^3 + axY - x^3 = 0 \tag{2.1}$$

as a series in x that will be valid near $x = 0$ and that will give the root $a \neq 0$ when $x = 0$. This computation can be found in the paragraph entitled *Exempla per Resolutionem Æquationum Affectarum*. The paragraph begins with what we now call "Newton's method" for finding roots, and the series solution is presented as an extension of that numerical method. We know of no earlier reference that could be considered to be a version of the implicit function theorem.

Newton refined his procedure in the 1670 manuscript *De Methodis Serierum et Fluxionum* (see the paragraph *De Affectarum Æquationum Reductione*), and the device constructed in this improvement is now known as the Newton polygon or Newton diagram.

To introduce the Newton polygon, we begin with an example.

Example 2.2.1 Consider the equation

$$y^3 + x^2 y^2 - xy + x^4 = 0 \tag{2.2}$$

near $x = 0$. The locus of points satisfying this equation is shown in Figure 2.1.

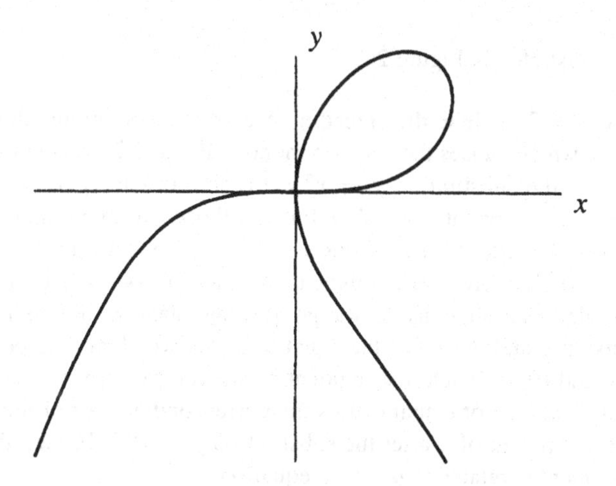

Figure 2.1. The Locus of Points Satisfying (2.2)

[3] This manuscript can be found together with its translation in Newton [NW 68].

Assume there is a solution of (2.2) of the form $y = y(x)$ that has its graph passing through the origin and that is defined at least for small values of x. In particular, we will have $y(0) = 0$.

The idea behind the Newton polygon is to make the further assumption that we can write

$$y(x) = x^{\alpha}\tilde{y}(x) \tag{2.3}$$

with $\tilde{y}(x)$ a continuous function that does not vanish when $x = 0$. The number α in (2.3) is a parameter which must be chosen appropriately. Newton's insight was that a value of α should be used if and only if its use allows $\tilde{y}(0)$ to be determined.

Substituting $y = x^{\alpha}\tilde{y}$ in (2.2), we obtain

$$x^{3\alpha}\tilde{y}^3 + x^{2\alpha+2}\tilde{y}^2 - x^{\alpha+1}\tilde{y} + x^4 = 0\,. \tag{2.4}$$

To be able to determine $\tilde{y}(0)$ from (2.4), there must be two or more monomials in (2.4) which have the same power of x and all other monomials must have a larger power of x.

For instance, if we set $\alpha = 3$, then (2.4) becomes

$$x^9\tilde{y}^3 + x^8\tilde{y}^2 - x^4\tilde{y} + x^4 = 0\,. \tag{2.5}$$

Dividing (2.5) by x^4, we obtain

$$x^5\tilde{y}^3 + x^4\tilde{y}^2 - \tilde{y} + 1 = 0\,. \tag{2.6}$$

Setting $x = 0$ in (2.6), we find $\tilde{y}(0) = 1$. This tells us that the locus of points satisfying (2.2) contains a curve approximated near $x = 0$ by

$$y = x^3\,.$$

This curve is illustrated in Figure 2.2. □

The choice $\alpha = 3$ made in the preceding example is not unique; this choice is merely the one which causes the last two monomials in (2.2) to contain the same power of x after the substitution $y = x^{\alpha}\tilde{y}$. In fact, for each pair of monomials in (2.2) there is an exponent α which will cause those two monomials to contain the same power of x after the substitution $y = x^{\alpha}\tilde{y}$. One convenient way to keep track of all these possibly useful values of α is as follows: For each nontrivial monomial in the equation, consider the point in the plane whose coordinates are the exponents on x and on y. For the equation (2.2), we obtain the points $(0, 3)$, $(2, 2)$, $(1, 1)$, and $(0, 4)$. Each line segment between a pair of these points can be identified with a choice of α that causes the corresponding pair of monomials to contain the same power of x after the substitution $y = x^{\alpha}\tilde{y}$. In fact, the slope m of the line segment is related to α by the equation

$$\alpha = -1/m\,,$$

as the reader should verify.

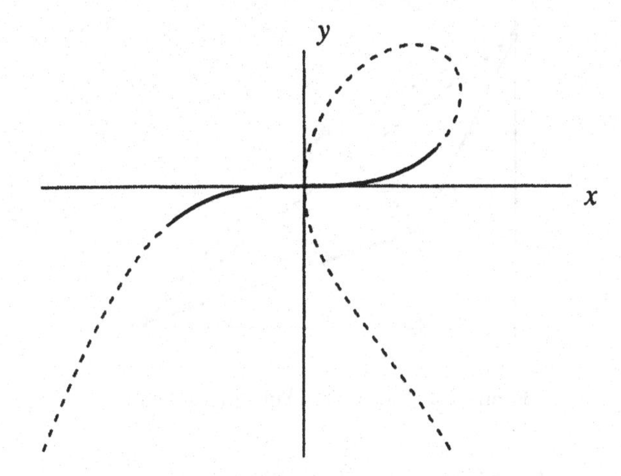

Figure 2.2. The Part of the Locus Approximated by $y = x^3$

Figure 2.3 shows all the line segments corresponding to pairs of monomials in (2.2). The associated values of α are 3, 1/2, 4/3, 2, 1, and -1. Only the first two choices corresponding to the substitutions $y = x^3\tilde{y}$ and $y = x^{1/2}\tilde{y}$, respectively, lead to curves that approximate part of the locus. Below we describe the geometric method used to decide which of the possible substitutions should be used.

The set of segments in Figure 2.3 encloses a convex region in the plane, namely, the convex hull of the set of points

$$\{(0,3),\ (2,2),\ (1,1),\ (0,4)\}.$$

Because there is no common power of x or y in (2.2), the convex region touches both the vertical and horizontal axes. The Newton polygon associated with (2.2) is the part of the boundary of the convex region that goes along the bottom left boundary of the region from the vertical axis to the horizontal axis (see Figure 2.4). Only values of α corresponding to segments in the Newton polygon

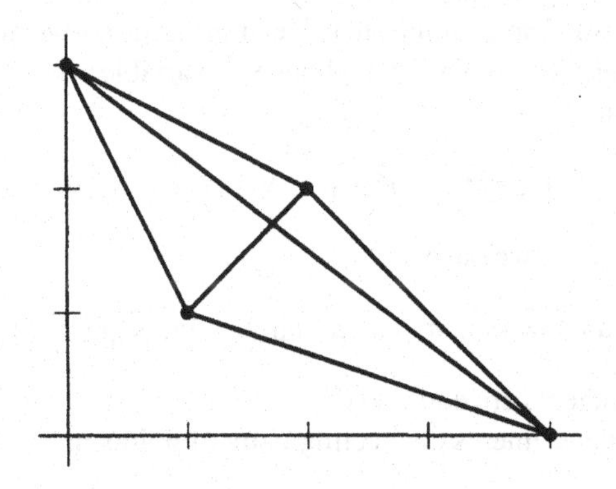

Figure 2.3. Segments Corresponding to Pairs of Monomials

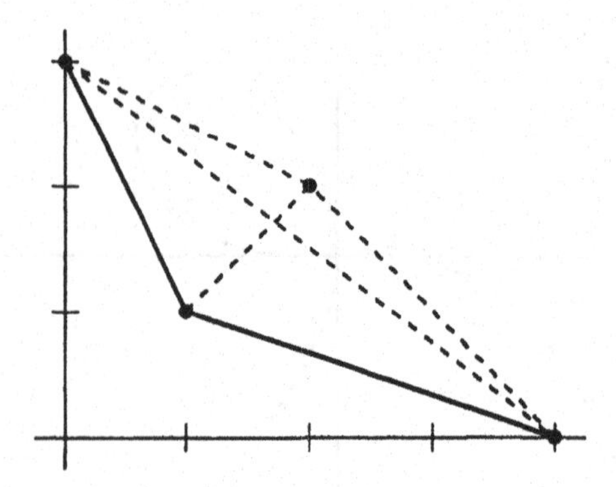

Figure 2.4. The Newton Polygon for (2.2)

allow a nonzero value of $\tilde{y}(0)$ to be determined. For example, $\alpha = -1$ corresponding to the segment from $(1, 1)$ to $(2, 2)$ is not part of the Newton polygon and the equation resulting from the substitution $y = x^{-1}\tilde{y}$ is

$$x^{-3}\tilde{y}^3 + x^{-2}\tilde{y}^2 + x^{-2}\tilde{y} + x^{-4} = 0$$

which cannot be satisfied by any function $\tilde{y}(x)$ that is continuous at $x = 0$.

General Construction of the Newton Polygon. The Newton polygon is used to determine the behavior of the locus of points satisfying a polynomial equation

$$P(x, y) = \sum_{n=0}^{N} \sum_{i+j=n} a_{i,j} x^i y^j \tag{2.7}$$

in a neighborhood of a point of the locus. By changing variables using a translation, we may assume that a point of the locus is (0,0). We may also assume that there is no common factor of x or y in the polynomial. Purists might wish to assume the irreducibility of P, but this is not necessary for the analysis that will follow.

Equation (2.1) was the example that Newton used, so we will use it here to illustrate the process. If we make the change of variable $y = Y - a$ in (2.1), we obtain the equation

$$y^3 + 3ay^2 + 4a^2y + axy + a^2x - x^3 = 0. \tag{2.8}$$

In the notation of (2.7), we have

$$a_{1,0} = a^2,\ a_{0,1} = 4a^2,\ a_{1,1} = a,\ a_{0,2} = 3a,\ a_{3,0} = -1,\ a_{0,3} = 1,$$

and all other coefficients are equal to 0.

The set of all line segments connecting pairs of points in

$$\{\, (i, j) \ :\ a_{i,j} \neq 0 \,\} \tag{2.9}$$

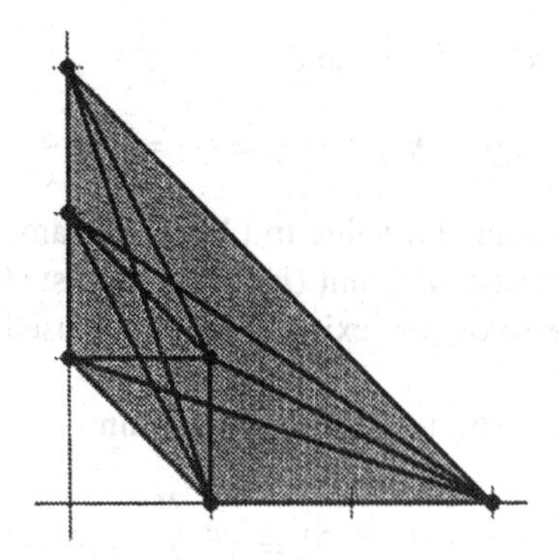

Figure 2.5. Constructing the Newton Polygon

encloses a convex set K. In fact, K is the convex hull of the set given in (2.9). The boundary of K, denoted ∂K, is a closed polygonal path in the first quadrant that intersects both axes. Of the two subpaths in ∂K with an endpoint in each axis, the *Newton polygon* is the one nearer the origin. This construction is illustrated in Figure 2.5 for the equation (2.8).

To appreciate the significance of the Newton polygon, let us rewrite the polynomial P in the form

$$P(x, y) = \sum_{j=0}^{M} A_j(x) x^{h_j} y^j, \tag{2.10}$$

where either $A_j \equiv 0$ or $A_j(0) \neq 0$ (if we were to have $A_j(0) = 0$, then a power of x would divide $A_j(x)$ and that power of x should have been factored out and included in x^{h_j}). The assumption that there is no common factor of x or y implies that A_0 is not the zero polynomial and that some $h_j = 0$. We have $h_0 \neq 0$, since $P(0, 0) = 0$.

Remark 2.2.2 Notice that if two or more of the h_j's in (2.10) were zero, then $P(0, y)$ would not be identically zero and thus would have at least one nonzero root r. Consequently, for small values of x there would be a root $y(x)$ of $P(x, y)$ near to r, that is, we can approximate one branch of the locus $P(x, y) = 0$ by the line $y = r$. Of course, we are interested in branches of the locus that pass through $(0, 0)$, rather than branches through $(0, r)$, but we will see that each segment of the Newton polygon allows us to reduce one branch through $(0, 0)$ to this simpler situation.

Any vertex of the Newton polygon must be of the form (h_j, j), so any line segment contained in the Newton polygon must contain two or more such points. We list those points as

$$(h_{j_1}, j_1), \ (h_{j_2}, j_2), \ldots, \ (h_{j_\sigma}, j_\sigma). \tag{2.11}$$

Letting $-1/\alpha$ be the slope of the line segment, we note that if we substitute

$$y(x) = x^\alpha \tilde{y}(x), \tag{2.12}$$

then we have $x^{h_{j_p}} y^{j_p} = x^{h_{j_p}+\alpha j_p} \tilde{y}^{j_p}$ and

$$h_{j_1} + \alpha j_1 = h_{j_2} + \alpha j_2 = \cdots = h_{j_\sigma} + \alpha j_\sigma \tag{2.13}$$

holds. Let β denote the common value in (2.13). For any j such that A_j is not identically zero and such that the point (h_j, j) is not listed in (2.11), we see that $h_j + \alpha j > \beta$, this because of convexity of the set K used to define the Newton polygon.

Thus, by making the substitution (2.12), we obtain

$$P(x, y) = \sum_{j=0}^{M} A_j(x) x^{h_j + \alpha j} \tilde{y}^j = x^\beta \sum_{j=0}^{M} A_j(x) x^{(h_j + \alpha j - \beta)} \tilde{y}^j .$$

Since the polynomial in $\tilde{y}$,

$$\widetilde{P}(x, \tilde{y}) = \sum_{j=0}^{M} A_j(x) x^{(h_j + \alpha j - \beta)} \tilde{y}^j ,$$

has two or more of the powers $h_j + \alpha j$ equal to zero, we find ourselves in the simpler situation discussed above in Remark 2.2.2—except that the terms that vanish when $x = 0$ now may involve positive fractional powers of x rather than only positive integral powers. Letting r be a nonzero root of $\widetilde{P}(0, \tilde{y}) = 0$, we conclude that a branch of the locus of $\widetilde{P}(x, \tilde{y}) = 0$ near $(0, r)$ can be approximated by the line $\tilde{y} = r$ and, thus, a branch of the locus of $P(x, y) = 0$ near $(0, 0)$ can be approximated by the curve $y = x^\alpha r$.

For the equation (2.8), there is only one segment in the Newton polygon and it has a slope of -1. Thus we substitute

$$y = x\tilde{y},$$

and, after eliminating a common factor of x, we find that

$$x^2 \tilde{y}^3 + 3ax\tilde{y}^2 + ax\tilde{y} - x^2 + 4a^2\tilde{y} + a^2 = 0. \tag{2.14}$$

The solution of (2.14) near $x = 0$ satisfies $\tilde{y} \approx -\frac{1}{4}$, so we conclude that $y \approx -\frac{1}{4}x$ and finally that

$$Y \approx a - \tfrac{1}{4}x.$$

Of course, in this case, the last approximation is the linear approximation readily obtained using calculus and the implicit function theorem.

2.3 Lagrange

Much of Lagrange's fame as a mathematician was owing to his successes in the study of celestial mechanics. He received numerous prizes for this work, begin-

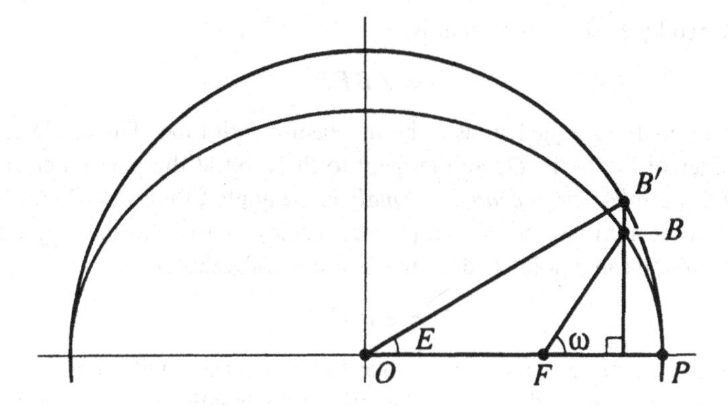

Figure 2.6. Orbital Parameters

ning with the 1764 award given by the Paris Academy of Sciences for his paper on the libration of the moon.[4]

A basic result in celestial mechanics is Kepler's equation

$$E = M + e\sin(E), \tag{2.15}$$

where M is the mean anomaly,[5] E is the eccentric anomaly, and e is the eccentricity of the orbit. We will describe these quantities in more detail later. For the moment, we note that M and e should be considered to be the quantities that can be measured and that e is assumed to be small. One of Lagrange's theorems, now called the Lagrange Inversion Theorem, gave a formula for the correction that must be made when, for some function $\psi(\cdot)$, $\psi(M)$ is replaced by $\psi(E)$. The correction takes the form of a power series in e. Thus, one can adjust for the difference between the mean anomaly and the eccentric anomaly. Since Lagrange was not sensitive to questions of convergence in the way we are today, his proof amounts to what we would call a "formal power series" argument.

Kepler's Equation. Kepler's (1571–1630) equation is

$$E = M + e\sin(E),$$

where M is the mean anomaly, E is the eccentric anomaly, and e is the eccentricity of the orbit. Figure 2.6 illustrates the true anomaly, ω, and the eccentric anomaly, E, of a body, B, moving in an elliptical orbit about a much more massive body at the focus, F, of the ellipse. The position of the body at a particular time is indicated by the point B. The *pericenter* of the orbit, P, is defined to be the point of nearest approach of the orbiting body to the focus F. The *true anomaly* is the

[4]The libration of the moon is an irregularity of its motion that allows approximately 59% of the moon's surface to be visible from the earth.

[5]In astronomy, the word "anomaly" refers to the angle between the direction to an orbiting body and the direction to its last perihelion.

angle formed by B, F, and P, that is,

$$\omega = \angle BFP. \tag{2.16}$$

The true anomaly is signed so as to be increasing with time. The circle centered at the center of the orbit, O, and tangent to the orbit at the pericenter is called the *auxillary circle*. The *eccentric anomaly* is the angle formed by P, O, and the point B' on the auxillary circle that projects orthogonally onto the major axis of the ellipse to the same point as does the orbiting body, that is,

$$E = \angle B'OP. \tag{2.17}$$

The eccentric anomaly is also signed so as to be increasing with time.

The *eccentricity*, e, of the orbit is the ratio of the length OF to the length OP. In Figure 2.6 the eccentricity is 0.6. The eccentricity of the earth's orbit about the sun is approximately 0.016, so, were the figure to be a representation of the earth and the sun, Figure 2.6 would be quite exaggerated.

The mean anomaly does not have a geometric description that can be illustrated readily in Figure 2.6. Rather, the *mean anomaly* is the angle

$$M = \angle \tilde{B}OP, \tag{2.18}$$

where $\tilde{B}$ is the location of a *hypothetical body* traveling around the auxillary circle with the same period of rotation as the orbiting body, but which is moving with constant speed. This hypothetical body is assumed to start from the pericenter at the same time (and in the same direction) as the actual orbiting body. The hypothetical and actual bodies will again be coincident at the far end of the major axis, and will coincide twice in each complete orbit. The mean anomaly is much more easily determined than the eccentric anomaly, but the eccentric anomaly is more relevant geometrically and physically.

Lagrange's Theorem. To state and prove Lagrange's theorem, we will need to use the language of and some results from complex analysis. The reader without the requisite background may simply take note of Lagrange's formula (2.21).

Theorem 2.3.1 (Lagrange Inversion Theorem [La 69]). *Let $\psi(z)$ and $\phi(z)$ be analytic on the open disc $D(a, r) \subset \mathbb{C}$ and continuous on the closed disc $\overline{D}(a, r)$. If t is of small enough modulus that*

$$|t\phi(z)| < |z - a| \tag{2.19}$$

holds for $z \in \partial D(a, r)$, then

$$\zeta = a + t\phi(\zeta) \tag{2.20}$$

has exactly one root in $D(a, r)$ and, if that root $\zeta = \zeta(t)$ is considered as a function of t, then we have

$$\psi(\zeta) = \psi(a) + \sum_{n=1}^{\infty} \frac{t^n}{n!} \left(\frac{d^{n-1}}{dz^{n-1}} [\psi'(z)\{\phi(z)\}^n] \right)\Bigg|_{z=a}. \tag{2.21}$$

We will give two proofs of Lagrange's theorem. The first proof uses the Cauchy theory from complex analysis. The second is a proof that is due to Laplace (1749–1827), and depends heavily on the chain rule of calculus.

We will need some classical results from complex analysis. The first of these classical results is the Cauchy integral formula (see Greene and Krantz [GK 97; page 48]).

Theorem 2.3.2 (Cauchy Integral Formula). *Suppose that U is an open set in $\mathbf{C}$ and that f is a holomorphic function on U. Let $z_0 \in U$ and let $r > 0$ be such that $\overline{D}(z_0, r) \subseteq U$. Then, for each $z \in \overline{D}(z_0, r)$, it holds that*

$$f(z) = \frac{1}{2\pi i} \oint_{\partial D(a,r)} \frac{f(\zeta)}{\zeta - z} \, d\zeta \, .$$

The second classical result we will need is Rouché's theorem. (See Greene and Krantz [GK 97; page 168ff.]).

Lemma 2.3.3 (Rouché's theorem). *Suppose that $f, g : U \to \mathbb{C}$ are analytic functions on an open set $U \subset \mathbb{C}$. If $\overline{D}(a, r) \subseteq U$ and if, for each $z \in \partial D(a, r)$,*

$$|f(z) - g(z)| < |f(z)| + |g(z)| \tag{2.22}$$

holds, then the number of zeros of f in $D(a, r)$ equals the number of zeros of g in $D(a, r)$, counting multiplicities.

Proof of Lagrange's Inversion Theorem 2.3.1 using the Cauchy Integral Formula. We will make the simplifying assumption that

$$\phi(z) \neq 0 \text{ in } \overline{D}(a, r) \, . \tag{2.23}$$

By Lemma 2.3.3, applied with $f(z) = z - a$ and with $g(z) = z - a - t\phi(z)$, we see that (2.20) has exactly one root ζ in $D(a, r)$.

Fix t and $\zeta = \zeta(t)$ satisfying (2.20). We set

$$\theta(z) = \frac{z - a}{\phi(z)} . \tag{2.24}$$

We have $\theta(\zeta) = t$.

Note that $\theta(z)$ is analytic on $D(a, r)$ and that $\theta(z) - \theta(\zeta)$ has its only zero at $z = \zeta$. We can write

$$\theta(z) = \theta(\zeta) + (z - \zeta)R(z),$$

where $R(z)$ is nonvanishing in $\overline{D}(a, r)$. Using the Cauchy integral formula and the fact that $R(\zeta) = \theta'(\zeta)$, we compute

$$\begin{aligned} \frac{1}{2\pi i} \oint_{\partial D(a,r)} \frac{\psi(z)\theta'(z)}{\theta(z) - \theta(\zeta)} \, dz &= \frac{1}{2\pi i} \oint_{\partial D(a,r)} \frac{\psi(z)\theta'(z)}{(z - \zeta)R(z)} \, dz \\ &= \frac{\psi(\zeta)\theta'(\zeta)}{R(\zeta)} = \psi(\zeta). \end{aligned}$$

The condition (2.19) is equivalent to $|\theta(\zeta)| < |\theta(z)|$, so we have

$$\frac{\theta'(z)}{\theta(z)-\theta(\zeta)} = \sum_{n=0}^{\infty} \frac{\theta'(z)[\theta(\zeta)]^n}{[\theta(z)]^{n+1}}. \tag{2.25}$$

Thus, we have

$$\begin{aligned}
\psi(\zeta) &= \frac{1}{2\pi i} \oint_{\partial D(a,r)} \frac{\psi(z)\theta'(z)}{\theta(z)-\theta(\zeta)}\, dz \\
&= \sum_{n=0}^{\infty} \frac{1}{2\pi i} \oint_{\partial D(a,r)} \frac{\psi(z)\theta'(z)[\theta(\zeta)]^n}{[\theta(z)]^{n+1}}\, dz \\
&= \sum_{n=0}^{\infty} [\theta(\zeta)]^n \frac{1}{2\pi i} \oint_{\partial D(a,r)} \frac{\psi(z)\theta'(z)}{[\theta(z)]^{n+1}}\, dz \\
&= \sum_{n=0}^{\infty} t^n \frac{1}{2\pi i} \oint_{\partial D(a,r)} \frac{\psi(z)\theta'(z)}{[\theta(z)]^{n+1}}\, dz.
\end{aligned}$$

Integration by parts gives us

$$\oint_{\partial D(a,r)} \frac{\psi(z)\theta'(z)}{[\theta(z)]^{n+1}}\, dz = \frac{1}{n} \oint_{\partial D(a,r)} \frac{\psi'(z)}{[\theta(z)]^n}\, dz.$$

So we have

$$\psi(\zeta) = \sum_{n=0}^{\infty} t^n \frac{1}{2n\pi i} \oint_{\partial D(a,r)} \frac{\psi'(z)}{[\theta(z)]^n}\, dz$$

Using equation (2.24), we have

$$\begin{aligned}
\psi(\zeta) &= \sum_{n=0}^{\infty} t^n \frac{1}{2n\pi i} \oint_{\partial D(a,r)} \frac{\psi'(z)}{[\theta(z)]^n}\, dz \\
&= \sum_{n=0}^{\infty} t^n \frac{1}{2n\pi i} \oint_{\partial D(a,r)} \frac{\psi'(z)[\phi(z)]^n}{(z-a)^n}\, dz \\
&= \sum_{n=0}^{\infty} t^n \frac{1}{n!} \frac{d^{n-1}}{da^{n-1}} \left(\psi'(a)[\phi(a)]^n\right).
\end{aligned}$$

□

We now present a second proof of Theorem 2.3.1 that is longer and less self-contained than the first. It utilizes some interesting new ideas, including the Schwarz–Pick lemma from complex variable theory, which we state next (see Greene and Krantz [GK 97; page 174]).

Lemma 2.3.4 (Schwarz–Pick Lemma). *Let h be analytic on the open unit disc in $\mathbb{C}$. If*

$$\begin{aligned}
|h(z)| &\le 1 \text{ for all } |z| < 1, \\
h(c) &= d,
\end{aligned}$$

then

$$|h'(c)| \leq \frac{1-|d|^2}{1-|c|^2}. \tag{2.26}$$

In case equality holds in (2.26), then h is of the form

$$h(z) = \omega \frac{z-\alpha}{1-\bar{\alpha}z},$$

for some complex numbers α *and* ω *with* $|\alpha| < 1$, $|\omega| = 1$, *and if, additionally,* $c = d = 0$, *then* $\alpha = 0$.

. In the proof given below for Theorem 2.3.1, we will also need to apply the maximum modulus theorem from complex analysis (see Greene and Krantz [GK 97; page 172]).

Theorem 2.3.5 (Maximum Modulus Theorem). *Let* $U \subseteq \mathbf{C}$ *be a bounded, open, connected set. Let* f *be a continuous function on* $\overline{U}$ *that is holomorphic on* U. *Then the maximum value of* $|f|$ *on* $\overline{U}$ *must occur on* ∂U.

Proof that ζ is analytic. We again begin by applying Rouché's Theorem 2.3.3 with $f(z) = z - a$ and $g(z) = t\phi(z)$ to see that (2.20) has exactly one root ζ in $D(a, r)$.

Now fix t and ζ, the corresponding root of (2.20). We will apply Lemma 2.3.4 to

$$h(z) = \frac{t}{r}\phi\left(a + r\frac{rz+\zeta-a}{r-z(\bar{\zeta}-\bar{a})}\right), \quad z \in D(0, 1). \tag{2.27}$$

The function

$$z \mapsto a + r\frac{rz+\zeta-a}{r-z(\bar{\zeta}-\bar{a})}$$

maps $D(0, 1)$ to $D(a, r)$, sending 0 to ζ. So $|h(z)| \leq |z - a|/r = 1$ persists for $z \in \partial D(0, 1)$, and the inequality holds on the interior of the unit disc by the maximum modulus theorem. Setting $c = 0$, $d = (\zeta - a)/r$ in Lemma 2.3.4, we conclude that

$$|h'(0)| \leq \frac{r^2 - |\zeta - a|^2}{r^2} \leq 1. \tag{2.28}$$

Now $|h'(0)| = 1$ implies both that $\zeta = a$ and that the case of equality has occurred in Lemma 2.3.4. So by the uniqueness part of Lemma 2.3.4, we can conclude that $h(z) = \omega z$, for some complex constant ω of modulus 1. It follows then that $\phi(z) = \frac{\omega}{t}(z - a)$, contradicting (2.19). Thus, we must have $|h'(0)| < 1$.

The inequality $|h'(0)| < 1$ implies that $1 - t\phi'(\zeta) \neq 0$, which is exactly the condition we need to apply the complex analytic form of the implicit function theorem (to be presented in the next section) to conclude that ζ is an analytic function of t. Indeed, for future purposes, we note that $1 - t\phi'(\zeta) \neq 0$ shows that ζ is an analytic function of *both* t and a. □

It remains to show that Lagrange's expansion (2.21) is valid.

Laplace's Proof of Lagrange's Expansion (2.21). Computing the partial derivatives of (2.20) with respect to a and t, we obtain

$$1 = [1 - t\phi'(\zeta)]\frac{\partial \zeta}{\partial a} \tag{2.29}$$

$$\phi(\zeta) = [1 - t\phi'(\zeta)]\frac{\partial \zeta}{\partial t}. \tag{2.30}$$

Writing

$$u(a, t) = \psi[\zeta(a, t)],$$

we find that

$$\frac{\partial u}{\partial t} = \phi(\zeta)\frac{\partial u}{\partial a}. \tag{2.31}$$

We will also need the general identity

$$\frac{\partial}{\partial a}\left[F(\zeta)\frac{\partial u}{\partial t}\right] = \frac{\partial}{\partial t}\left[F(\zeta)\frac{\partial u}{\partial a}\right] \tag{2.32}$$

that holds for any differentiable F. To see that (2.32) holds, we compute, on the one hand,

$$\begin{aligned} \frac{\partial}{\partial a}\left[F(\zeta)\frac{\partial u}{\partial t}\right] &= F'(\zeta)\frac{\partial \zeta}{\partial a}\frac{\partial u}{\partial t} + F(\zeta)\frac{\partial^2 u}{\partial a \partial t} \\ &= F'(\zeta)\frac{\partial \zeta}{\partial a}\psi'(\zeta)\frac{\partial \zeta}{\partial t} + F(\zeta)\frac{\partial^2 u}{\partial a \partial t} \end{aligned}$$

and, on the other hand,

$$\begin{aligned} \frac{\partial}{\partial t}\left[F(\zeta)\frac{\partial u}{\partial a}\right] &= F'(\zeta)\frac{\partial \zeta}{\partial t}\frac{\partial u}{\partial a} + F(\zeta)\frac{\partial^2 u}{\partial a \partial t} \\ &= F'(\zeta)\frac{\partial \zeta}{\partial t}\psi'(\zeta)\frac{\partial \zeta}{\partial a} + F(\zeta)\frac{\partial^2 u}{\partial a \partial t}. \end{aligned}$$

To verify the Lagrange expansion, we need to show that the coefficients in the Taylor series for $u(a, t)$, considered as a function of t, are as given in (2.21); that is, we need to show that

$$\frac{\partial^n u}{\partial t^n} = \frac{\partial^{n-1}}{\partial a^{n-1}}\left[\phi(\zeta)^n \frac{\partial u}{\partial a}\right] \tag{2.33}$$

holds for $n = 1, 2, \ldots$. This will be proved by induction on n. Note that (2.33) holds when $n = 1$ by (2.31).

To see that the inductive step is true, suppose that (2.33) holds for the positive integer n. We compute

$$\begin{aligned}
\frac{\partial^{n+1}u}{\partial t^{n+1}} &= \frac{\partial}{\partial t}\left(\frac{\partial^{n-1}}{\partial a^{n-1}}\left[\phi(\zeta)^n\frac{\partial u}{\partial a}\right]\right) \\
&= \frac{\partial^{n-1}}{\partial a^{n-1}}\left(\frac{\partial}{\partial t}\left[\phi(\zeta)^n\frac{\partial u}{\partial a}\right]\right) \\
&= \frac{\partial^{n-1}}{\partial a^{n-1}}\left(\frac{\partial}{\partial a}\left[\phi(\zeta)^n\frac{\partial u}{\partial t}\right]\right) && \text{by (2.32)} \\
&= \frac{\partial^{n}}{\partial a^{n}}\left[\phi(\zeta)^n\frac{\partial u}{\partial t}\right] \\
&= \frac{\partial^{n}}{\partial a^{n}}\left[\phi(\zeta)^{n+1}\frac{\partial u}{\partial a}\right] && \text{by (2.31),}
\end{aligned}$$

which verifies that (2.33) holds for $n + 1$. □

2.4 Cauchy

As mentioned in Section 2.1, Cauchy (1789–1857) is credited with the first rigorous form of the implicit function theorem. The next theorem, in the context of holomorphic functions, proves the existence of an implicitly defined function under the now standard hypotheses and also gives an integral representation for the function; the argument used in the proof is due to Cauchy.[6] As in the statement and proof of Lagrange's theorem, techniques from complex analysis will be used in this section. The reader without background in that area may wish to skip this section. The final result in this section applies to formal power series.

Theorem 2.4.1 *Suppose that $F(x, y)$ is holomorphic in the bidisc $D(x_0, R_1) \times D(y_0, R_2) \subseteq \mathbf{C}^2$ and write*

$$D_2F = \frac{\partial F}{\partial y}. \tag{2.34}$$

If

$$F(x_0, y_0) = 0 \text{ and } D_2F(x_0, y_0) \neq 0, \tag{2.35}$$

then there is a disc $D(x_0, r_0)$ and a unique holomorphic function $f(x)$ defined on $D(x_0, r_0)$ with $f(x_0) = y_0$ and such that

$$F(x, f(x)) = 0 \tag{2.36}$$

[6] Cauchy [Ca 16; page 74ff.]

holds for $x \in D(x_0, r_0)$. Moreover, that function $f(x)$ is represented by

$$f(x) = \frac{1}{2\pi i} \oint_C y \frac{D_2 F(x, y)}{F(x, y)} dy, \tag{2.37}$$

where $C = \partial D(y_0, r_1)$ is a suitably chosen circle.

Proof. By the hypotheses (2.35) we see that, as a function of y, $F(x_0, y)$ has a simple zero at $y = y_0$. It follows that there exists $0 < r_1 < R_2$ such that

$$F(x_0, y) \neq 0 \text{ holds for } 0 < |y - y_0| \leq r_1. \tag{2.38}$$

In particular, we have

$$0 < \inf\{|F(x_0, y)| : |y - y_0| = r_1\}. \tag{2.39}$$

Since $F(x, y)$ approaches $F(x_0, y)$, uniformly in y, as x approaches x_0, and because of (2.39), we can select a number $0 < r_0 < R_1$ such that

$$|F(x, y) - F(x_0, y)| < |F(x_0, y)| \text{ holds for } |x - x_0| \leq r_0,\ |y - y_0| = r_1. \tag{2.40}$$

Now, by Rouché's theorem (i.e., Lemma 2.3.3) and (2.40), for each fixed x with $|x - x_0| \leq r_0$, the functions $F(x, y)$ and $F(x_0, y)$ have the same number of zeros in the disc $D(x_0, r_1)$, and since $F(x_0, y)$ has exactly one zero, it follows that $F(x, y)$ also has exactly one zero, which we may denote by $f(x)$.

It is evident that, for fixed $x \in D(x_0, r_0)$, the residue of

$$y \frac{D_2 F(x, y)}{F(x, y)}$$

as a function of y at the point $y = f(x)$ is just $f(x)$, so the representation (2.37) holds. The fact that $f(x)$ is a holomorphic function of x then follows by differentiating (2.37), with respect to x, under the integral sign. □

Remark 2.4.2 The proof given above can also be adapted to the situation in which $F(x_0, y)$ has a zero of multiplicity $m > 1$ at y_0. In this case, for each fixed $x \in D(x_0, r_1)$, it is the sum of the zeros of $F(x, \cdot)$ in $D(y_0, r_1)$ that is given by the right-hand side of (2.37); of course the zeros must be counted according to their multiplicities. In fact, Cauchy dealt extensively with this form of the result.

Cauchy also gave a proof of the implicit function theorem by means of majorants.[7] The proof by the method of majorants is equally applicable to real analytic functions and holomorphic functions, since only the convergence of power series is at issue. A complete treatment of the real analytic implicit function theorem,

[7] What is now known as the "method of majorants" was called the *calcul des limites* by Cauchy.

together with its connections to the complex holomorphic implicit function theorem, appears in [KP 92].

The method of majorants is also the key tool in the proof of the Cauchy–Kowalewsky theorem (Sonja Kowalewsky: 1853–1891) on the existence of solutions of certain partial differential equations (see Courant and Hilbert [CH 62; Chapter 1, Section 7] or Krantz and Parks [KP 92; Sections 1.7 and 1.10]).

We will need a result from several complex variables which allows us to bound the coefficients in a convergent power series (see Krantz [Kr 92; Section 2.3]); this result is a consequence of the *Cauchy estimates* in several variables.

Lemma 2.4.3 *If*

$$f(x_1, x_2, \ldots, x_n) = \sum_{j_1, j_2, \ldots, j_n=0}^{\infty} \gamma_{j_1 j_2 \ldots j_n} x_1^{j_1} x_2^{j_2} \ldots x_n^{j_n}$$

is absolutely convergent for $|x_1| \le R_1$, $|x_2| \le R_2, \ldots, |x_n| \le R_n$ *and if*

$$M = \sup\{|f(x)| : x \in \bar{D}(0, R_1) \times \bar{D}(0, R_2) \times \cdots \times \bar{D}(0, R_n)\},$$

then

$$|\gamma_{j_1 j_2 \ldots j_n}| \le \frac{M}{R_1^{j_1} R_2^{j_2} \ldots R_n^{j_n}}$$

holds for $j_1, j_2, \ldots, j_n \in \{0, 1, \ldots\}$.

Theorem 2.4.4 *Suppose the power series*

$$F(x, y) = \sum_{j,k=0}^{\infty} a_{jk} x^j y^k \tag{2.41}$$

is absolutely convergent for $|x| \le R_1$, $|y| \le R_2$. *If*

$$a_{00} = 0 \text{ and } a_{01} \ne 0, \tag{2.42}$$

then there exist $r_0 > 0$ *and a power series*

$$f(x) = \sum_{j=1}^{\infty} c_j x^j \tag{2.43}$$

such that (2.43) is absolutely convergent for $|x| \le r_0$ *and*

$$F(x, f(x)) = 0. \tag{2.44}$$

Proof. It will be no loss of generality to assume $a_{01} = 1$, so that (2.41) takes the form

$$F(x, y) = y + \sum_{j=1}^{\infty} (a_{j0} + a_{j1} y) x^j + \sum_{j=0}^{\infty} \sum_{k=2}^{\infty} a_{jk} x^j y^k. \tag{2.45}$$

Introducing the notation $b_{jk} = -a_{jk}$, we can rewrite the equation $F(x, y) = 0$ as

$$y = \sum_{j=1}^{\infty}(b_{j0} + b_{j1}y)x^j + \sum_{j=0}^{\infty}\sum_{k=2}^{\infty} b_{jk}x^j y^k \tag{2.46}$$

or $y = B(x, y)$, where

$$B(x, y) = \sum_{j=1}^{\infty}(b_{j0} + b_{j1}y)x^j + \sum_{j=0}^{\infty}\sum_{k=2}^{\infty} b_{jk}x^j y^k. \tag{2.47}$$

Substituting $y = f(x)$ into (2.46) with $f(x)$ given by (2.43) we obtain

$$\sum_{j=1}^{\infty} c_j x^j = \sum_{j=1}^{\infty} b_{j0}x^j + \sum_{j=1}^{\infty}\sum_{k=1}^{\infty} b_{j1}c_k x^{j+k} + \sum_{j=0}^{\infty}\sum_{k=2}^{\infty} b_{jk}x^j \left(\sum_{\ell=1}^{\infty} c_\ell x^\ell\right)^k. \tag{2.48}$$

If all the series in (2.48) are ultimately shown to be absolutely convergent, then the order of summation can be freely rearranged. Assuming absolute convergence, we can equate like powers of x on the left-hand and right-hand sides of (2.48) and obtain the following sequence of recurrence relations that must hold:

$$\begin{aligned}
c_1 &= b_{10}, \\
c_2 &= b_{20} + b_{11}c_1 + b_{02}(c_1)^2 \\
c_3 &= b_{30} + b_{21}c_1 + b_{12}c_2 + b_{12}(c_1)^2 + b_{03}(c_1)^3 + b_{11}c_2 + 2b_{02}c_1c_2, \\
&\vdots \\
c_J &= b_{J0} + b_{(J-1)1}c_1 + \cdots + b_{1J}c_J \\
&\quad + \sum \frac{k!}{k_1!k_2!\ldots k_\rho!} b_{jk}(c_1)^{k_1}(c_2)^{k_2}\ldots(c_\rho)^{k_\rho}, \\
&\vdots
\end{aligned} \tag{2.49}$$

where the last summation extends over $j \in \{0, 1, \ldots\}$, $k \in \{2, 3, \ldots\}$, $\rho \in \{1, 2, \ldots\}$, and $k_1, k_2, \ldots, k_\rho \in \{0, 1, \ldots\}$ such that

$$j + k_1 + 2k_2 + \cdots + \rho k_\rho = J.$$

While the recurrence relations (2.49) uniquely determine the coefficients c_j in the power series for the implicit function, it is also necessary to show that (2.43) is convergent. The easiest way to obtain the needed estimates is by using the method of majorants, which we describe next. Consider two power series in the same number of variables:

$$\Phi(x_1, x_2, \ldots, x_\rho) = \sum_{j_1, j_2, \ldots, j_\rho = 0}^{\infty} \phi_{j_1, j_2, \ldots, j_\rho} x_1^{j_1} x_2^{j_2} \ldots x_\rho^{j_\rho}, \tag{2.50}$$

$$\Psi(x_1, x_2, \ldots, x_\rho) = \sum_{j_1, j_2, \ldots, j_\rho = 0}^{\infty} \psi_{j_1, j_2, \ldots, j_\rho} x_1^{j_1} x_2^{j_2} \ldots x_\rho^{j_\rho}. \tag{2.51}$$

We say that $\Psi(x_1, x_2, \ldots, x_\rho)$ is a *majorant* of $\Phi(x_1, x_2, \ldots, x_\rho)$ if

$$|\Phi_{j_1,j_2,\ldots,j_\rho}| \leq \Psi_{j_1,j_2,\ldots,j_\rho} \tag{2.52}$$

holds for all $j_1, j_2, \ldots, j_\rho$.

Because all the coefficients $k!/(k_1!\,k_2!\ldots k_\rho!)$ in (2.49) are non-negative, if

$$G(x, y) = \sum_{j,k=0}^{\infty} g_{jk} x^j y^k,$$

with $g_{00} = g_{01} = 0$, is a majorant of

$$B(x, y) = \sum_{j=1}^{\infty} (b_{j0} + b_{j1} y) x^j + \sum_{j=0}^{\infty} \sum_{k=2}^{\infty} b_{jk} x^j y^k$$

and if

$$h(x) = \sum_{j=1}^{\infty} h_j x^j \tag{2.53}$$

solves

$$h(x) = G[x, h(x)], \tag{2.54}$$

then $h(x)$ will be a majorant of $f(x)$. Consequently, if the series (2.53) for $h(x)$ is convergent, then the series (2.43) is convergent and its radius of convergence is at least as large as the radius of convergence for (2.53).

We take

$$\begin{aligned} G(x, y) &= M\left[(1 - x/R_1)^{-1}(1 - y/R_2)^{-1} - 1 - y/R_2\right] \\ &= M\left[-1 - \frac{y}{R_2} + \sum_{j,k=0}^{\infty} \frac{x^j y^k}{(R_1)^j (R_2)^k}\right] \\ &= M\left[\sum_{j}^{\infty} \left(1 + \frac{y}{R_2}\right) \frac{x^j}{(R_1)^j} + \sum_{j=0}^{\infty} \sum_{k=2}^{\infty} \frac{x^j y^k}{(R_1)^j (R_2)^k}\right] \end{aligned}$$

where

$$M = \sup\{|B(x, y)| : x \in \bar{D}(0, R_1),\ y \in \bar{D}(0, R_2)\}.$$

We see that G is a majorant of B by Lemma 2.4.3. For this choice of majorant, (2.54) can be solved explicitly and the solution is clearly holomorphic at $x = 0$. If fact, $y = h(x)$ is easily seen to be the solution of the quadratic equation

$$\frac{M + R_2}{M(R_2)^2} y^2 - \frac{1}{M} y - 1 + \left(1 - \frac{x}{R_1}\right)^{-1} = 0. \tag{2.55}$$

□

Remark 2.4.5

1. By using a smaller majorant a better estimate on the radius of convergence of f can be obtained (see Hille [Hi 59; pages 272–273]).
2. The recurrence relations (2.49) constitute the implicit function theorem for formal power series. For completeness we state the formal power series theorem below.
3. In Section 6.1 we treat the N-dimensional version of the ideas formulated in Theorem 2.4.4.

Theorem 2.4.6 *Suppose the formal power series*

$$\mathcal{F}(x, y) = \sum_{j,k=0}^{\infty} a_{jk} x^j y^k \tag{2.56}$$

satisfies

$$a_{00} = 0 \text{ and } a_{01} \neq 0. \tag{2.57}$$

Then there exists a unique formal power series $f(x) = \sum c_j x^j$ *satisfying* $\mathcal{F}(x, f(x)) = 0$ *and that series has its coefficients given by the following recurrence relations:*

$$\begin{aligned}
c_0 &= 0,\\
c_1 &= -\frac{a_{10}}{a_{01}},\\
c_2 &= -\frac{a_{20}}{a_{01}} - \frac{a_{11}}{a_{01}}c_1 - \frac{a_{02}}{a_{01}}(c_1)^2,\\
c_3 &= -\frac{a_{30}}{a_{01}} - \frac{a_{21}}{a_{01}}c_1 - \frac{a_{12}}{a_{01}}c_2 - \frac{a_{12}}{a_{01}}(c_1)^2 - \frac{a_{03}}{a_{01}}(c_1)^3 - \frac{a_{11}}{a_{01}}c_2 - 2\frac{a_{02}}{a_{01}}c_1c_2,\\
&\vdots\\
c_J &= -\frac{a_{J0}}{a_{01}} - \frac{a_{(J-1)1}}{a_{01}}c_1 - \cdots - \frac{a_{1J}}{a_{01}}c_J\\
&\quad -\sum \frac{k!}{k_1!\,k_2!\ldots k_p!}\frac{a_{jk}}{a_{01}}(c_1)^{k_1}(c_2)^{k_2}\ldots(c_p)^{k_p},
\end{aligned} \tag{2.58}$$

where the summation in (2.58) extends over

$$j \in \{0, 1, \ldots\},\ k \in \{2, 3, \ldots\},\ p \in \{1, 2, \ldots\},$$

and

$$k_1, k_2, \ldots, k_p \in \{0, 1, \ldots\}$$

such that

$$j + k_1 + 2k_2 + \cdots + pk_p = J.$$

Prior to this text, the history of the implicit function theorem was not generally well known. Most of the material that we present comes from primary sources. The resulting tapestry illustrates a striking synergy among the different contributors to the idea of what we now think of as a basic theorem of the calculus.

We hope that the genesis presented here will give the reader some context and motivation for what follows.

3

Basic Ideas

3.1 Introduction

In order to make this book a convenient reference, we shall endeavor to make it locally self-contained. With this thought in mind, we shall begin by presenting a very classical treatment of the implicit function theorem in Euclidean space.

There are two basic points of view in this classical setting. The first is to prove the implicit function theorem as an exercise in calculus (the Taylor expansion, the mean-value theorem, and estimates on derivatives) and the second is to use the Contraction Mapping Fixed Point Principle from elementary functional analysis to get a quick, soft, and easy proof of the implicit function theorem.

We will give two proofs of the elementary calculus type. The first illustrates the original proof of the real-variable implicit function theorem. The result is obtained for just one dependent variable, but an arbitrary number of independent variables; then the general result is obtained by induction on the number of dependent variables. Our second proof based on elementary calculus looks directly at the linear approximation of the mapping provided by calculus. This second proof reveals the inner workings of the theorem.

The functional analysis approach to the implicit function theorem will provide us with our third proof of the result. This method of proof has the disadvantage of being more abstract. On the other hand, that very abstraction allows the argument to be applied in other settings, thus automatically yielding variants of the implicit function theorem in categories other than C^k.

3.2 The Inductive Proof of the Implicit Function Theorem

In this section, we present a proof of the implicit function theorem that is essentially the one given by Dini in the 1870s. The proof proceeds by induction on the number of dependent variables. The basis of the induction—the implicit function theorem for one dependent variable, one equation, and any number of independent variables—is relatively easy. The induction step is accomplished by distinguishing one equation and one dependent variable to which the base case can be applied, but with all the other dependent variables treated as if they were independent. The resulting implicitly defined function then is substituted in the other equations, thereby reducing the number of dependent variables and equations by one.

First, we state and prove the implicit function theorem for one dependent variable and one equation, but any number of independent variables. The proof relies on the intermediate value theorem and the use of a nonvanishing derivative to insure monotonicity. The hypotheses can be weakened a bit as in Young [Yo 09a] while still maintaining the same general method of proof.

Theorem 3.2.1 *If $W \subseteq \mathbf{R}^m$ is open, $F : W \to \mathbf{R}$ is continuously differentiable, and $p = (p', q) \in \mathbf{R}^{m-1} \times \mathbf{R}$, where $p \in W$ is a point for which*

$$F(p) = 0 \quad \textit{and} \quad \frac{\partial F}{\partial x_m}(p) \neq 0, \tag{3.1}$$

then there exists an open set $W' \subset \mathbf{R}^{m-1}$ with $p' \in W'$ and a unique continuously differentiable function $\psi : W' \to \mathbf{R}$ such that $q = \psi(p')$ and

$$F[x', \psi(x')] = 0 \tag{3.2}$$

holds for $x' \in W'$.

Proof. Without loss of generality, we may assume that

$$\frac{\partial F}{\partial x_m}(p) > 0. \tag{3.3}$$

By the continuity of $\partial F/\partial x_m$, and by passing to a smaller neighborhood W of p if necessary, but without changing notation, we may assume that

$$\frac{\partial F}{\partial x_m}(x) > 0 \tag{3.4}$$

holds for all $x \in W$.

Since (by (3.4)) $F(p', \cdot)$ is an increasing function in an interval about q, we can find $q_1 < q < q_2$ so that

$$F(p', q_1) < 0 < F(p', q_2) \tag{3.5}$$

holds. Using the continuity of F, we can find a neighborhood W' of p' so that $W' \times [q_1, q_2] \subseteq W$ and

$$F(x', q_1) < 0 < F(x', q_2) \tag{3.6}$$

holds for $x' \in W'$. It follows from the intermediate value theorem that for each $x' \in W'$ there is a number y with $q_1 < y < q_2$ so that $F(x', y) = 0$ and by (3.4) that number y is unique. We let that y be $\psi(x')$ and note that the uniqueness of this value y also implies that ψ is continuous.

To complete the proof, we need to show that

$$\frac{\partial \psi}{\partial x_j} = -\frac{\partial F}{\partial x_j} \Big/ \frac{\partial F}{\partial x_m} \tag{3.7}$$

holds, for $j = 1, 2, \ldots, m-1$. We do this with a very direct argument. Specifically, fixing the point $x' \in W'$ and setting $y = \psi(x')$, we have

$$F(x' + se_j, y + t) - F(x', y) = s\frac{\partial F}{\partial x_j}(x', y) + t\frac{\partial F}{\partial x_m}(x', y) + \epsilon\sqrt{s^2 + t^2} \tag{3.8}$$

where $\epsilon = \epsilon(s, t)$ approaches 0 as $\sqrt{s^2 + t^2}$ approaches 0. Now, taking $t = \psi(x' + se_j) - \psi(x')$ in (3.8), we find that

$$t\,\frac{\partial F}{\partial x_m}(x', y) = -s\,\frac{\partial F}{\partial x_j}(x', y) - \epsilon\sqrt{s^2 + t^2}\,. \tag{3.9}$$

So we have

$$|t|\,\left|\frac{\partial F}{\partial x_m}(x', y)\right| \le |s|\,\left|\frac{\partial F}{\partial x_j}(x', y)\right| + |\epsilon|\,|s| + |\epsilon|\,|t|. \tag{3.10}$$

For small enough choice of $|s|$, we have $|\epsilon| \le \frac{1}{2}\,|\partial F/\partial x_m(x', y)|$ and $|\epsilon| \le 2\,|\partial F/\partial x_j(x', y)|$, so that

$$|t| \le 6\,|s|\,\left|\frac{\partial F}{\partial x_j}(x', y)\right| \Big/ \left|\frac{\partial F}{\partial x_m}(x', y)\right| \tag{3.11}$$

holds. Thus we see that

$$\left|\frac{\psi(x' + se_j) - \psi(x')}{s} + \left[\frac{\partial F}{\partial x_j}(x', y)\right]\left[\frac{\partial F}{\partial x_m}(x', y)\right]^{-1}\right|$$

$$\le |\epsilon|\left(1 + 6\,\left|\frac{\partial F}{\partial x_j}(x', y)\right|\,\left|\frac{\partial F}{\partial x_m}(x', y)\right|^{-1}\right). \tag{3.12}$$

Taking the limit as $s \to 0$ in (3.12), we see that $\partial\psi/\partial x_j$ exists at x' and is given by the formula in (3.7). The continuous differentiability of ψ follows. □

Our proof of the general implicit function theorem will be simplified notationally if we use the following lemma from linear algebra.

Lemma 3.2.2 *Let A be an $n \times n$ real matrix. Then there exists an invertible real matrix U such that UA is upper triangular.*

Proof. The matrix A can be reduced to row echelon form by a sequence of elementary row operations. A square matrix in row echelon form is necessarily upper triangular, and each elementary row operation can be accomplished via left multiplication by an invertible matrix. The result follows. □

Now we set up the notation that we will use in the general theorem.

Notation 3.2.3 We suppose that we are given a set of equations

$$f_i(x_1, x_2, \ldots, x_\ell; y_1, y_2, \ldots, y_n) = 0, \quad i = 1, 2, \ldots, n, \tag{3.13}$$

where the functions $f_1, f_2, \ldots, f_n$ are continuously differentiable. We will assume that $(p; q) = (p_1, p_2, \ldots, p_\ell; q_1, q_2, \ldots, q_n)$ is a point such that all the equations (3.13) hold and at which we have

$$\det \begin{pmatrix} \frac{\partial f_1}{\partial y_1} & \frac{\partial f_1}{\partial y_2} & \cdots & \frac{\partial f_1}{\partial y_n} \\ \frac{\partial f_2}{\partial y_1} & \frac{\partial f_2}{\partial y_2} & \cdots & \frac{\partial f_2}{\partial y_n} \\ \vdots & \vdots & & \vdots \\ \frac{\partial f_n}{\partial y_1} & \frac{\partial f_n}{\partial y_2} & \cdots & \frac{\partial f_n}{\partial y_n} \end{pmatrix} \neq 0. \tag{3.14}$$

We can think of the functions $f_i(p; \cdot)$ as giving a mapping $F : \mathbf{R}^n \to \mathbf{R}^n$ defined by

$$y \mapsto F(y) = \Big(f_1(p; y), f_2(p; y), \ldots, f_n(p; y)\Big). \tag{3.15}$$

Lemma 3.2.2, applied with $A = DF(q)$, provides us with a linear transformation (namely left multiplication by the invertible matrix U from Lemma 3.2.2) that can be composed with this function F to give us a new function

$$y \mapsto \widehat{F}(y) = \Big(\widehat{f_1}(p; y), \widehat{f_2}(p; y), \ldots, \widehat{f_n}(p; y)\Big)$$

such that

$$\frac{\partial \widehat{f_i}}{\partial y_j}(p; q) = 0 \quad \text{whenever } i > j.$$

It is more convenient to simply assume, without changing notation, that we have

$$\frac{\partial f_i}{\partial y_j}(p; q) = 0 \quad \text{whenever } i > j. \tag{3.16}$$

After this preliminary modification, we have

$$\begin{pmatrix} \frac{\partial f_1}{\partial y_1} & \frac{\partial f_1}{\partial y_2} & \cdots & \frac{\partial f_1}{\partial y_n} \\ \frac{\partial f_2}{\partial y_1} & \frac{\partial f_2}{\partial y_2} & \cdots & \frac{\partial f_2}{\partial y_n} \\ \vdots & \vdots & & \vdots \\ \frac{\partial f_n}{\partial y_1} & \frac{\partial f_n}{\partial y_2} & \cdots & \frac{\partial f_n}{\partial y_n} \end{pmatrix} = \begin{pmatrix} \frac{\partial f_1}{\partial y_1} & \frac{\partial f_1}{\partial y_2} & \cdots & \frac{\partial f_1}{\partial y_n} \\ 0 & \frac{\partial f_2}{\partial y_2} & \cdots & \frac{\partial f_2}{\partial y_n} \\ \vdots & \vdots & & \vdots \\ 0 & 0 & \cdots & \frac{\partial f_n}{\partial y_n} \end{pmatrix} \tag{3.17}$$

and, consequently,

$$\det \begin{pmatrix} \frac{\partial f_1}{\partial y_1} & \frac{\partial f_1}{\partial y_2} & \cdots & \frac{\partial f_1}{\partial y_n} \\ \frac{\partial f_2}{\partial y_1} & \frac{\partial f_2}{\partial y_2} & \cdots & \frac{\partial f_2}{\partial y_n} \\ \vdots & \vdots & & \vdots \\ \frac{\partial f_n}{\partial y_1} & \frac{\partial f_n}{\partial y_2} & \cdots & \frac{\partial f_n}{\partial y_n} \end{pmatrix} = \prod_{i=1}^{n} \frac{\partial f_i}{\partial y_i} \neq 0 \tag{3.18}$$

at the point $(p; q)$.

Theorem 3.2.4 *There exists a neighborhood $U \subset \mathbf{R}^\ell$ of p and a set of continuously differentiable functions $\phi_j : U \to \mathbf{R}$, $j = 1, 2, \ldots, n$, such that $\phi_j(p) = q_j$, $j = 1, 2, \ldots, n$, and*

$$f_i[x; \phi_1(x), \phi_2(x), \ldots, \phi_n(x)] = 0, \quad i = 1, 2, \ldots, n, \tag{3.19}$$

hold for $x \in U$.

Proof. We argue by induction on n. The case $n = 1$ is of course Theorem 3.2.1.

Suppose now that $n > 1$ and that the theorem is true with n replaced by $n - 1$. We assume that we have done the preliminary simplification as in Notation 3.2.3. By (3.18), we have

$$\frac{\partial f_n}{\partial y_n}(p; q) \neq 0. \tag{3.20}$$

Let us introduce the notation $y' = (y_1, y_2, \ldots, y_{n-1})$; then Theorem 3.2.1 is applicable to the equation

$$f_n(x; y'; y_n) = 0 \tag{3.21}$$

at the point $(p; q'; q_n)$, where we are treating the variables $x_1, x_2, \ldots, x_\ell$ and $y_1, y_2, \ldots, y_{n-1}$ as independent and only the variable y_n as dependent. Thus, by

Theorem 3.2.1, there is a neighborhood $V \subset \mathbf{R}^{\ell+n-1}$ of $(p; q')$ and a continuously differentiable function $\psi : V \to \mathbf{R}$ such that $\psi(p; q') = q_n$ and

$$f_n[x; y'; \psi(x; y')] = 0 \tag{3.22}$$

holds for $(x; y') \in V$.

Notice that, if (3.22) is differentiated with respect to y_j, $1 \leq j \leq n-1$, then we find that

$$\frac{\partial f_n}{\partial y_j} + \frac{\partial f_n}{\partial y_n}\frac{\partial \psi}{\partial y_j} = 0. \tag{3.23}$$

Evaluating (3.23) at $x = p$, $y' = q'$, and using (3.16) and (3.20), we see that

$$\frac{\partial \psi}{\partial y_j}(p; q') = 0 \tag{3.24}$$

holds for $j = 1, 2, \ldots, n-1$.

Now, for each $i = 1, 2, \ldots, n-1$, define the function h_i by setting

$$h_i(x_1, x_2, \ldots, x_\ell; y_1, y_2, \ldots, y_{n-1}) = f_i[x; y', \psi(x; y')]. \tag{3.25}$$

Consider the system of equations

$$h_i(x_1, x_2, \ldots, x_\ell; y_1, y_2, \ldots, y_{n-1}) = 0, \quad i = 1, 2, \ldots, n-1. \tag{3.26}$$

For $j = 1, 2, \ldots, n-1$, by (3.24), we have

$$\frac{\partial h_i}{\partial y_j}(p; q') = \frac{\partial f_i}{\partial y_j}(p; q'; q_n) + \frac{\partial f_i}{\partial y_n}(p; q'; q_n)\frac{\partial \psi}{\partial y_j}(p; q') = \frac{\partial f_i}{\partial y_j}(p; q) \tag{3.27}$$

and, accordingly (by (3.18)),

$$\det \begin{pmatrix} \frac{\partial h_1}{\partial y_1} & \frac{\partial h_1}{\partial y_2} & \cdots & \frac{\partial h_1}{\partial y_{n-1}} \\ \frac{\partial h_2}{\partial y_1} & \frac{\partial h_2}{\partial y_2} & \cdots & \frac{\partial h_2}{\partial y_{n-1}} \\ \vdots & \vdots & & \vdots \\ \frac{\partial h_{n-1}}{\partial y_1} & \frac{\partial h_{n-1}}{\partial y_2} & \cdots & \frac{\partial h_{n-1}}{\partial y_{n-1}} \end{pmatrix}$$

$$= \det \begin{pmatrix} \frac{\partial f_1}{\partial y_1} & \frac{\partial f_1}{\partial y_2} & \cdots & \frac{\partial f_1}{\partial y_{n-1}} \\ \frac{\partial f_2}{\partial y_1} & \frac{\partial f_2}{\partial y_2} & \cdots & \frac{\partial f_2}{\partial y_{n-1}} \\ \vdots & \vdots & & \vdots \\ \frac{\partial f_{n-1}}{\partial y_1} & \frac{\partial f_{n-1}}{\partial y_2} & \cdots & \frac{\partial f_{n-1}}{\partial y_{n-1}} \end{pmatrix} \neq 0. \tag{3.28}$$

Here the partial derivatives in the left-hand determinant are evaluated at $(p; q')$ and those in the right-hand determinant are evaluated at $(p; q)$.

By induction, there exist a neighborhood U' of p in $\mathbf{R}^\ell$ and continuously differentiable functions $\phi_j : U' \to \mathbf{R}$ such that $\phi_j(p) = q_j$ and

$$h_i[x; \phi_1(x), \phi_2(x), \ldots, \phi_{n-1}(x)] = 0, \quad i = 1, 2, \ldots, n-1, \tag{3.29}$$

hold for $x \in U'$.

Set $\Phi(x) = (x, \phi_1(x), \ldots, \phi_{n-1}(x))$ and

$$U = U' \cap \Phi^{-1}(V) \tag{3.30}$$

and define $\phi_n : U \to \mathbf{R}$ by setting

$$\phi_n(x) = \psi[x; \phi_1(x), \phi_2(x), \ldots, \phi_{n-1}(x)]. \tag{3.31}$$

By the definition of the h_i (equation (3.25)) we see that the desired equations (3.19) hold. □

3.3 The Classical Approach to the Implicit Function Theorem

The development of the implicit function theorem can be traced by looking at some older textbooks. Initially the implicit function theorem is treated as a theorem about a function of two real variables. The nondegeneracy condition in the variable "to be solved for" is traditionally formulated as a monotonicity hypothesis (this can be seen for example in Hobson [Ho 57; Section 38]). This approach provided for a simple statement and proof of the theorem, but ignored the behavior of the Jacobian determinant which is the crux of the matter. It was, of course, Dini who realized how to formulate the result in the context of several variables and who used the Jacobian determinant to provide the correct nondegeneracy hypothesis. Dini's proof was inductive (see Section 3.2) and (one may feel) not as revealing as our more modern proofs.

It is appropriate to begin our detailed discussion of the implicit and inverse function theorems with a review of the Jacobian matrix, the Jacobian determinant, and their role in the calculus. Let U, V be open subsets of $\mathbf{R}^N$ and let $G : U \to V$ be a C^1 mapping. We write $G(x) = (g_1(x), \ldots, g_N(x))$. If $p \in U$, then the *Jacobian matrix* of G at p is

$$DG(p) \equiv \begin{pmatrix} \dfrac{\partial g_1}{\partial x_1}(p) & \dfrac{\partial g_1}{\partial x_2}(p) & \cdots & \dfrac{\partial g_1}{\partial x_N}(p) \\ \dfrac{\partial g_2}{\partial x_1}(p) & \dfrac{\partial g_2}{\partial x_2}(p) & \cdots & \dfrac{\partial g_2}{\partial x_N}(p) \\ \vdots & \vdots & & \vdots \\ \dfrac{\partial g_N}{\partial x_1}(p) & \dfrac{\partial g_N}{\partial x_2}(p) & \cdots & \dfrac{\partial g_N}{\partial x_N}(p) \end{pmatrix}.$$

The Jacobian matrix plays the same role in calculus of several variables as the first derivative (a 1×1 matrix) does in the calculus of one variable. In particular, the C^1 mapping G is well approximated near p by its Jacobian matrix $DG(p)$. More precisely, the change in G, that is, $G(p+\mathbf{v}) - G(p)$, is well approximated by the vector $\mathbf{v}$ left-multiplied by $DG(p)$. In order to recall the calculus concept of "differential," we use the letter "D" to denote this matrix.

In the context of the inverse function theorem, we consider the *Jacobian determinant*, which is

$$\det DG(p) = \det \begin{pmatrix} \frac{\partial g_1}{\partial x_1}(p) & \frac{\partial g_1}{\partial x_2}(p) & \cdots & \frac{\partial g_1}{\partial x_N}(p) \\ \frac{\partial g_2}{\partial x_1}(p) & \frac{\partial g_2}{\partial x_2}(p) & \cdots & \frac{\partial g_2}{\partial x_N}(p) \\ \vdots & \vdots & & \vdots \\ \frac{\partial g_N}{\partial x_1}(p) & \frac{\partial g_N}{\partial x_2}(p) & \cdots & \frac{\partial g_N}{\partial x_N}(p) \end{pmatrix}$$

The basic result, as we shall see below, is that when $\det DG(p) \neq 0$, then the restriction of the mapping G to a small neighborhood of p is invertible.

The Jacobian matrix represents the aggregate of information about the first-order behavior of a function near a point. As such, we apply the matrix to a (not necessarily unit) vector $\mathbf{v}$ to obtain information about a directional derivative in the direction of that vector. Then we denote the directional derivative of the function G in the direction $\mathbf{v}$ at the point p by

$$\langle DG(p), \mathbf{v} \rangle .$$

This notation is to be read as the matrix $DG(p)$ applied to the vector $\mathbf{v}$ using ordinary matrix multiplication.

We note that there are many different notations for the fundamental idea of Jacobian. A number of sources denote the Jacobian matrix of the mapping G by $\text{Jac}\, G$. The Jacobian determinant is then $\det \text{Jac}\, G$. Other references will denote the Jacobian determinant by $|\text{Jac}\, G|$. Still other works use the word "Jacobian" to *mean* the Jacobian determinant. In the present book, we use "Jacobian" to mean the Jacobian matrix (denoted DG) and "Jacobian determinant" to mean the determinant of the Jacobian matrix (denoted $\det DG$). An excellent reference for the concepts of Jacobian and Jacobian determinant, from the point of view of the calculus, is [Fl 77].

For the implicit function theorem, we do not consider equidimensional mappings. Therefore we are forced to look at the Jacobian *in the variables for which we wish to solve*. A useful notation in this context involves the components of a function, say

$$\Phi(x) = \Phi(x_1, \ldots, x_N) \equiv (\phi_1(x_1, \ldots, x_N), \ldots \phi_M(x_1, \ldots, x_N)),$$

a choice of some M arguments of the function, say

$$x_{i_1}, x_{i_2}, \ldots, x_{i_M},$$

and expresses the appropriate Jacobian determinant in the form

$$\frac{\partial(\phi_1, \ldots, \phi_M)}{\partial(x_{i_1}, \ldots, x_{i_M})} = \det \begin{pmatrix} \frac{\partial \phi_1}{\partial x_{i_1}} & \frac{\partial \phi_1}{\partial x_{i_2}} & \cdots & \frac{\partial \phi_1}{\partial x_{i_M}} \\ \frac{\partial \phi_2}{\partial x_{i_1}} & \frac{\partial \phi_2}{\partial x_{i_2}} & \cdots & \frac{\partial \phi_2}{\partial x_{i_M}} \\ \vdots & \vdots & & \vdots \\ \frac{\partial \phi_M}{\partial x_{i_1}} & \frac{\partial \phi_M}{\partial x_{i_2}} & \cdots & \frac{\partial \phi_M}{\partial x_{i_M}} \end{pmatrix}.$$

A standard formulation of the implicit function theorem is this:

Theorem 3.3.1 (The Implicit Function Theorem). *Let*

$$\Phi(x) = \Phi(x_1, \ldots, x_N) \equiv (\phi_1(x_1, \ldots, x_N), \ldots, \phi_M(x_1, \ldots, x_N))$$

be a mapping of class C^k, $k \geq 1$, defined on an open set $U \subseteq \mathbf{R}^N$ and taking values in $\mathbf{R}^M$. We assume that $1 \leq M < N$. Set $Q = N - M$.

Let $x^0 = (x_1^0, \ldots, x_N^0)$ be a fixed point of U. Of course we let $x = (x_1, \ldots, x_N)$ be any point of U. Set

$$x_a = (x_1, \ldots, x_Q) \quad \text{and} \quad x_a^0 = (x_1^0, \ldots, x_Q^0).$$

We suppose that

$$\frac{\partial(\phi_1, \ldots, \phi_M)}{\partial(x_{Q+1}, \ldots, x_N)}(x^0) \neq 0. \tag{3.32}$$

Then there exists a neighborhood $\widetilde{U}$ of x^0, and open set $W \subseteq \mathbf{R}^Q$ containing x_a^0, and functions $f_1, \ldots, f_M$ of class C^k on W such that

$$\Phi(x_1, \ldots, x_Q, f_1(x_a), \ldots, f_M(x_a)) = 0 \quad \text{for every } x_a \in W. \tag{3.33}$$

Furthermore $f_1, \ldots, f_M$ are the unique functions satisfying

$$\{x \in \widetilde{U} : \Phi(x) = 0\} = \{x \in \widetilde{U} : x_a \in W,\ x_{Q+\ell} = f_\ell(x_a) \text{ for } \ell = 1, \ldots, M\}.$$

As a companion result, we now formulate (in consistent notation) the inverse function theorem.

Theorem 3.3.2 (The Inverse Function Theorem). *Let $\widehat{W} \subseteq \mathbf{R}^Q$ be an open set and let $G : \widehat{W} \to \mathbf{R}^Q$ be a mapping of class C^k, $k \geq 1$.*

Let x^0 be a fixed point of $\widehat{W}$, and assume that $\det DG(x^0) \neq 0$. Then there exists a neighborhood $W \subseteq \widehat{W}$ of x^0 such that

1. *The restriction* $G\big|_W$ *is univalent;*

2. *The set* $V = G(W)$ *is open;*

3. *The inverse* G^{-1} *of* $G\big|_W$ *is of class* C^k.

We now present a classical proof of our two main theorems. More precisely, we shall prove that the inverse function theorem implies the Implicit Function Theorem (the converse is trivial). Then we shall prove the inverse function theorem.

By "classical" here, we mean an argument that uses only calculus and elementary estimates. The reader is encouraged to compare and contrast this proof with the Banach space proof that is given in Section 3.4. After we present the classical ideas, we shall see what consequences may be drawn from both arguments.

Proof that the Inverse Function Theorem Implies the Implicit Function Theorem

Our mapping Φ is at least C^1; hence the Jacobian determinant (3.32) of Φ is continuous. So there is a neighborhood $\widetilde{U}_0$ of x^0 on which this Jacobian does not vanish.

Let us consider the transformation

$$G : \widetilde{U}_0 \longrightarrow \mathbf{R}^N$$

given by

$$G(x) = (x_1, \ldots, x_Q, \phi_1(x), \ldots, \phi_M(x)) .$$

Then of course G is a mapping of class C^k, just as is Φ. Its Jacobian matrix is

$$\begin{pmatrix} 1 & 0 & \cdots & 0 & 0 & \cdots & 0 \\ 0 & 1 & \cdots & 0 & 0 & \cdots & 0 \\ \vdots & \vdots & & \vdots & \vdots & & \vdots \\ 0 & 0 & \cdots & 1 & 0 & \cdots & 0 \\ \partial\phi_1/\partial x_1 & & \cdots & \partial\phi_1/\partial x_Q & \partial\phi_1/\partial x_{Q+1} & \cdots & \partial\phi_1/\partial x_N \\ \vdots & & & & & & \vdots \\ \partial\phi_M/\partial x_1 & & \cdots & \partial\phi_M/\partial x_Q & \partial\phi_M/\partial x_{Q+1} & \cdots & \partial\phi_M/\partial x_N \end{pmatrix} .$$

Obviously the determinant of this matrix is just the determinant of the $M \times M$ block in the lower-right-hand corner. Since this is merely the Jacobian in (3.32), we see that $\det DG(x) \neq 0$ for every $x \in \widetilde{U}_0$.

By the inverse function theorem, we may now conclude that there is a neighborhood W of x^0 such that $G(W)$ is an open set and the restriction $G\big|_W$ has an inverse G^{-1} of class C^k.

Now let us write $(x_a, 0) = (x_1, \ldots, x_Q, 0, \ldots, 0)$. We set

$$R = \{x_a : (x_a, 0) \in G(W)\}.$$

Since $G(W)$ is an open subset of $\mathbf{R}^N$, R is also open. Now, for every $x_a \in R$, we let

$$f_\ell(x_a) = g_{Q+\ell}(x_a, 0), \qquad \ell = 1, \ldots, M.$$

For $x \in W$, $\Phi(x) = 0$ if and only if $x_a \in R$ and $G(x) = (x_a, 0)$. Since $G|_W$ and G^{-1} are inverses, $G(x) = (x_a, 0)$ if and only if $x = G^{-1}(x_a, 0)$. This completes the proof. □

Proof of the Inverse Function Theorem

This proof is considerable work, and we divide it into three steps. The argument follows the one that appears in Fleming [Fl 77].

Step 1: *The mapping G is locally univalent.*
Fix an arbitrary point $\hat{t} \in \widehat{W}$. Let $\mathbf{h}$ denote the inverse matrix of DG, and let $\|\mathbf{h}\|$ denote the norm of $\mathbf{h}$ considered as a linear operator on $\mathbf{R}^Q$. Define

$$c = \frac{1}{\|\mathbf{h}(\hat{t})\|}.$$

Let $\mathbf{L}$ denote the Jacobian matrix DG of G at $\hat{t}$, and set $\widetilde{G}(t) = G(t) - \mathbf{L}$. Then, for $s, t \in \widehat{W}$,

$$G(s) - G(t) = \mathbf{L}(s) - \mathbf{L}(t) + [\widetilde{G}(s) - \widetilde{G}(t)].$$

But

$$|\widetilde{G}(s) - \widetilde{G}(t)|/|s - t| \to 0 \qquad \text{as } s, t \to 0.$$

Therefore, for $\epsilon > 0$,

$$|G(s) - G(t)| \geq |\mathbf{L}(s) - \mathbf{L}(t)| - \epsilon \cdot |s - t|$$

provided that s is close to t. But, plainly,

$$|\mathbf{L}(s) - \mathbf{L}(t)| \geq c \cdot |s - t|. \tag{3.34}$$

This shows that

$$|G(s) - G(t)| \geq (c - \epsilon)|s - t|.$$

If we take $\epsilon = c/2$, then we find that

$$|G(s) - G(t)| \geq [c/2] \cdot |s - t|.$$

Thus $G(s) = G(t)$ implies that $s = t$, and we see that the mapping G is locally one-to-one on $\widehat{W}$.

Step 2: *The set $G(\widehat{W})$ is open.*
Set $V = G(\widehat{W})$. Let x^* be any point of V. We show that x^* has a neighborhood that lies in V.

By Step 1 we may choose $t^* \in W$ such that $G(t^*) = x^*$, where W is a neighborhood of t^* on which G is univalent. Let W^* be an open ball about t^* whose closure lies in W, and let S denote the boundary of W^*. Univalence now implies that $x^* \notin G(S)$. Of course the image under G of S is compact. Let

$$\sigma^* = \frac{1}{2}\mathrm{dist}(x^*, G(S)) .$$

Let $V^* = B(x^*, \sigma^*)$.

Now fix an arbitrary point $x \in V^*$. Then, for every $t \in S$,

$$2\sigma^* \leq |x^* - G(t)| \leq |x^* - x| + |x - G(t)| .$$

Since $|x^* - x| < \sigma^*$, we see that

$$\sigma^* < |x - G(t)|$$

for every $t \in S$.

For $t \in \widehat{W}$, define

$$F(t) = |x - G(t)|^2 = \sum_{j=1}^{Q}[x_j - g_j(t)]^2 ,$$

where the g_j are the components of G. Then F is of class C^k and must have a minimum on the compact Q-ball $\overline{W^*}$. But

$$F(t^*) = |x - x^*|^2 < [\sigma^*]^2$$

and

$$F(t) > [\sigma^*]^2 \quad \text{for every } t \in S .$$

It follows that the minimum value of F on $\overline{W^*}$ is less than $[\sigma^*]^2$ and hence must occur at some (interior) point $\widetilde{t}$ of W^*. Hence the partial derivatives of F at $\widetilde{t}$ must be 0.

Since

$$\frac{\partial}{\partial t_k}F(t) = -2\sum_{j=1}^{Q}(x_j - g_j(t))\frac{\partial}{\partial t_k}g_j(t)$$

($\partial/\partial t_k$ is used because x is fixed and t is variable), we have (setting $c_j = x_j - g_j(\widetilde{t})$)

$$0 = \sum_{j=1}^{Q} c_j \frac{\partial}{\partial x_t} g_j(\widetilde{t}) , \quad \text{each } k .$$

Of course $\det DG(\widetilde{t}) \neq 0$, hence the column vectors

$$\left(\frac{\partial g_1}{\partial t_j}(\widetilde{t}), \ldots, \frac{\partial g_Q}{\partial t_j}(\widetilde{t})\right), \qquad j = 1, \ldots, Q,$$

are linearly independent. We conclude that $c_j = 0$, $j = 1, \ldots, Q$, and hence that $x = G(\widetilde{t})$.

We have proved that if $x \in V^*$, then $x = G(\widetilde{t})$ for some $\widetilde{t} \in W^*$, or $x \in G(W^*)$. Thus $V^* \subseteq G(W^*) \subseteq V$, so $V = G(W)$ is open.

Notation. For use in the remainder of the proof, we fix a neighborhood W of x^0 on which G is univalent and on which $\det DG$ never equals 0.

Step 3: *The mapping $(G|_W)^{-1}$ is of class C^1.*
First notice that $(G|_W)^{-1}$ exists, by the local univalence established in Step 1. Let $x^* \in V$ and $t^* = (G|_W)^{-1}(x^*)$ as in the preceding step. Set $\mathbf{L}^* = DG(t^*)$. We now show that G^{-1} is differentiable at x^* and that $DG^{-1}(x^*) = (\mathbf{L}^*)^{-1}$.

Set $\widetilde{c} = 1/\|[\mathbf{L}^*]^{-1}\|$. For any $\epsilon > 0$, there is a ball $\widetilde{B} \equiv B(t^*, r^*) \subseteq W$ such that

$$|G(t) - G(t^*) - \mathbf{L}^*(t - t^*)| \leq \frac{\epsilon \widetilde{c} c}{2}|t - t^*| \tag{3.35}$$

for every $t \in \widetilde{B}$. Here the constant c is as in (3.34) in Step 1.

By Step 2, there is a neighborhood $B^* \equiv B(x^*, s^*)$ such that $B^* \subseteq G(\widetilde{B})$. Let $x \in B^*$. Then $x = G(t)$ for some $t \in \widetilde{B}$. Since $x^* = G(t^*)$, we find from Step 1 that

$$\frac{c}{2} \cdot |t - t^*| \leq |x - x^*|. \tag{3.36}$$

Further, since $t = G^{-1}(x)$ and $t^* = G^{-1}(x^*)$, we see that

$$\mathbf{L}^*[G^{-1}(x) - G^{-1}(x^*) - [\mathbf{L}^*]^{-1}(x - x^*)] = -[G(t) - G(t^*) - \mathbf{L}^*(t - t^*)].$$

Since $\widetilde{c}|\omega| \leq \mathbf{L}^*(\omega)|$ for every $\omega \in \mathbf{R}^Q$, we find that

$$\widetilde{c}|G^{-1}(x) - G^{-1}(x^*) - [\mathbf{L}^*]^{-1}(x - x^*)| \leq |G(t) - G(t^*) - \mathbf{L}^*(x - x^*)|.$$

Thus (3.35) and (3.36) yield, for every $x \in B^*$, that

$$|G^{-1}(x) - G^{-1}(x^*) - [\mathbf{L}^*]^{-1}(x - x^*)| \leq \epsilon|x - x^*|.$$

We see therefore that G^{-1} is differentiable at x^* and $DG^{-1}(x^*) = \mathbf{L}^{-1}$.

To summarize, the mapping G^{-1} is a differentiable function and

$$DG^{-1}(x) = \left[DG(G^{-1}(x))\right]^{-1} \tag{3.37}$$

for every $x \in V$. Of course it then follows that G^{-1} is continuous. Since each $\partial g_i/\partial x_j$ is a continuous function, the composition

$$\frac{\partial g_i}{\partial x_j} \circ G$$

is also continuous. It follows then from (3.37) (and Cramer's rule) that the partial derivatives

$$\frac{\partial [G^{-1}]_i}{\partial x_j}$$

are all continuous. As a result, the mapping G^{-1} is of class C^1.

If now G is of class C^m, then each $\partial g_i/\partial x_j$ is of class C^{m-1} and therefore $[\partial g_i/\partial x_j] \circ G^{-1}$ is of class C^{m-1}. As a result, $[\partial/\partial x_j][G^{-1}]_i$ is of class C^{m-1} and so G^{-1} is of class C^m. Inductively, we find that if G of class C^k, then so is G^{-1}. That completes the proof of Step 3.

Steps 1, 2, and 3 taken together complete our proof of the inverse function theorem, and therefore of the implicit function theorem. □

3.4 The Contraction Mapping Fixed Point Principle

Let X be a complete metric space with metric ρ. A mapping $F : X \to X$ is called a *contraction* if there is a constant $0 < c < 1$ such that

$$\rho(F(x), F(y)) \leq c \cdot \rho(x, y)$$

for all $x, y \in X$. The fact that $c < 1$ tells us that the image of a set under F is *contracted*—the points in $F(X)$ are closer together than are their pre-images in X. The basic theorem about contraction mappings is as follows.

Theorem 3.4.1 (Contraction Mapping Fixed Point Principle). *Let $F : X \to X$ be a contraction of the complete metric space X. Then F has a unique fixed point. That is, there is a unique point $x_0 \in X$ such that $F(x_0) = x_0$.*

Proof. Let $P \in X$ be any point. We define a sequence inductively by

$$\begin{aligned} x_1 &= P \\ x_2 &= F(x_1) \\ &\vdots \\ x_j &= F(x_{j-1}). \end{aligned}$$

We claim that $\{x_j\}$ is a Cauchy sequence in X. Suppose for the moment that this claim has been proved. Then, because X is complete, there is a limit point x_0 of the sequence. Furthermore,

$$F(x_0) = F(\lim_{j\to\infty} x_j) = \lim_{j\to\infty} F(x_j) = \lim_{j\to\infty} x_{j+1} = x_0.$$

So x_0 is certainly a fixed point for the mapping F. If $\widetilde{x_0}$ were another fixed point, then we would have

$$\rho(x_0, \widetilde{x_0}) = \rho(F(x_0), F(\widetilde{x_0})) \le c \cdot \rho(x_0, \widetilde{x_0}).$$

Since $0 < c < 1$, the only possible conclusion is that $\rho(x_0, \widetilde{x_0}) = 0$ or $x_0 = \widetilde{x_0}$. That establishes the existence and uniqueness of x_0. It remains to prove the claim.

We calculate, for $j \ge 1$, that

$$\begin{aligned}
\rho(x_j, x_{j-1}) &= \rho(F(x_{j-1}), F(x_{j-2})) \le c \cdot \rho(x_{j-1}, x_{j-2}) \\
&= c \cdot \rho(F(x_{j-2}), F(x_{j-3})) \le c^2 \cdot \rho(x_{j-2}, x_{j-3}) \\
&\le \cdots \le c^{j-1} \cdot \rho(x_1, x_0).
\end{aligned}$$

As a result,

$$\begin{aligned}
\rho(x_{j+k}, x_j) &\le \rho(x_{j+k}, x_{j+k-1}) + \rho(x_{j+k-1}, x_{j+k-2}) + \cdots + \rho(x_{j+1}, x_j) \\
&\le \left[c^{j+k-1} + c^{j+k-2} + \cdots + c^j\right]\rho(x_1, x_0) \\
&\le c^j \cdot \frac{1}{1-c}\rho(x_1, x_0)\,.
\end{aligned}$$

In particular, if $\epsilon > 0$ and if j is large enough then, regardless of the value of $k \ge 1$,

$$\rho(x_{j+k}, x_j) < \epsilon.$$

Thus the sequence $\{x_j\}$ is Cauchy, and the claim is proved. □

In fact we shall need in practice a slight variant of the contraction mapping fixed point theorem. We now state and establish this result.

Proposition 3.4.2 *Let $\overline{B} = \overline{B}(p, r)$ be a closed ball in a complete metric space X. Suppose that $H : \overline{B} \to X$ is a contraction (with contraction constant c, $0 < c < 1$) such that $\rho(H(p), p) \le (1 - c)r$. Then H has a unique fixed point in $\overline{B}$.*

Proof. If $x \in \overline{B}$ is any point, then

$$\begin{aligned}
\rho(H(x), p) &\le \rho(H(x), H(p)) + \rho(H(p), p) \\
&\le c \cdot \rho(x, p) + (1 - c)r \\
&\le cr + (1 - c)r = r\,.
\end{aligned}$$

This inequality verifies that the image of H lies in $\overline{B}$, hence $H : \overline{B} \to \overline{B}$. Theorem 3.4.1 now applies with X replaced by $\overline{B}$. The proof is therefore complete.

□

Proposition 3.4.3 *Let $B = B(p, r)$ be an open ball in a complete metric space X. Assume that $H : B \to X$ is a contraction (with contraction constant c, $0 < c < 1$) such that $\rho(H(p), p) < (1 - c)r$. Then H has a unique fixed point in B.*

Proof. Simply restrict H to a slightly smaller closed ball $\overline{D}$ that is concentric with and contained in B. Apply Proposition 3.4.2. □

Proposition 3.4.4 *Suppose that H is a contraction (with contraction constant c, $0 < c < 1$) on the complete metric space X. Let $x \in X$, and suppose that $\rho(H(x), x) = d$. Then the distance from x to the fixed point p (guaranteed to exist by Theorem 3.4.1) is at most $d/(1 - c)$.*

Proof: Let $\overline{B}$ denote the closed ball in X with center x and radius $r = d/(1 - c)$. Apply Proposition 3.4.2 to the restriction of H to $\overline{B}$. Thus the fixed point lies in $\overline{B}$, and we are done. □

Proposition 3.4.5 *Let X be a complete metric space, and S any metric space. Suppose that $H : S \times X \to X$. Assume that $H(s, x)$ is a contraction in X uniformly over $s \in S$ (with uniform contraction constant c, $0 < c < 1$), that is,*

$$\rho(H(s, x), H(s, y)) \leq c \cdot \rho(x, y)$$

for all $s \in S$ and all $x, y \in X$. Further, assume that H is continuous in s for each fixed $x \in X$. For each $s \in S$, let $p_s \in X$ be the unique fixed point satisfying $H(s, p_s) = p_s$. Then the map $s \mapsto p_s$ is a continuous function of s.

Proof. Choose $t \in S$. Let $\epsilon > 0$. By the continuity of H in the first variable, choose $\delta > 0$ so that if $\rho(s, t) < \delta$, then $\rho(H(s, p_t), H(t, p_t)) < \epsilon$. Since $H(t, p_t) = p_t$, this last inequality says that the contraction with parameter value s moves p_t a distance at most ϵ. Thus $\rho(p_s, p_t) \leq \epsilon/(1 - c)$ by Proposition 3.4.4.

That is, $\rho(s, t) < \delta$ implies that $\rho(p_s, p_t) < \epsilon/(1 - c)$. We conclude that the mapping $s \to p_s$ is continuous at $t \in S$. □

Now we draw together Propositions 3.4.3 and 3.4.5 into a single result which will be the one that is used to establish the implicit function theorem (Theorem 3.4.10). The reader should bear in mind that this next theorem is just a slight variant of the basic contraction mapping fixed point result given by Theorem 3.4.1. Because we have taken the trouble to do some elementary preliminary manipulations with the contraction mapping theorem, our proof of the implicit function theorem will therefore be short and elegant.

Theorem 3.4.6 *Let X be a complete metric space, and S any metric space. Let $B = B(p, r)$ be an open ball in X. Let H be a mapping from $S \times B$ to X which is a contraction in X uniformly over $s \in S$ (with uniform contraction constant c, $0 < c < 1$); further, suppose that H is continuous in the s variable for each fixed value* *of $x \in X$. Finally assume that, for each $s \in S$, we know that $\rho(H(s, p), p) < (1 - c)r$.*

Then, for each $s \in S$, there is a unique $p_s \in B$ such that $H(s, p_s) = p_s$; furthermore, the mapping $s \mapsto p_s$ is continuous from S to B.

Proof. The result is immediate from Propositions 3.4.3 and 3.4.5. □

Now the implicit function theorem, formulated in a manner that lends itself to proof by the contraction mapping fixed point principle, will be our next major result. We first introduce some definitions and notation.

Notation 3.4.7 (Landau) Fix a in the extended reals, that is, $a \in \mathbf{R} \cup \{\pm\infty\}$. Suppose g is a real-valued function on $\mathbf{R} \cup \{\pm\infty\}$ that does not vanish in a neighborhood of a. For a real-valued function f defined in a punctured neighborhood of a, we say f is *little "o" of g as $x \to a$* and write

$$f(x) = o(g(x)) \text{ as } x \to a$$

in case

$$\lim_{x \to a} \frac{f(x)}{g(x)} = 0 .$$

Definition 3.4.8 Let X, Z be normed linear spaces. Let $x \in X$, and suppose that U is a neighborhood of x in X. A mapping $F : U \to Z$ is *differentiable* at x if there is a linear operator T from X to Z such that

$$F(x + \xi) - F(x) = T(\xi) + o(\|\xi\|)$$

for all small $\xi \in X$.

Of course this definition, commonly known as the "Fréchet definition" of derivative, simply says that the function F may be well approximated at x by the linear map T. It is straightforward to check that, if T exists, then it is unique. We call T the *derivative* of F at x. The reader may verify as an exercise that this definition is consistent with the standard one for differentiability in finite dimensions.

Definition 3.4.9 If X, Y, Z are Banach spaces and $F : X \times Y \to Z$ is a mapping then, for each fixed $x_0 \in X$, we may consider the differentiability of the mapping

$$Y \ni y \mapsto F(x_0, y).$$

If the derivative exists at a point $y_0 \in Y$, then we denote it by $d_2F(x_0, y_0)$.

Now we have the implicit function theorem, formulated so that it can be proved with the contraction mapping fixed point principle:

Theorem 3.4.10 *Let X, Y, Z be Banach spaces. Let $U \times V$ be an open subset of $X \times Y$. Suppose that $G : U \times V \to Z$ is continuous and has the property that d_2G exists and is continuous at each point of $U \times V$.*

Assume that the point $(x, y) \in X \times Y$ has the property that $G(x, y) = 0$ and that $d_2G(x, y)$ is invertible.

Then there are open balls $M = B_X(x, r)$ and $N = B_Y(y, s)$ such that, for each $\zeta \in M$, there is a unique $\eta \in N$ satisfying $G(\zeta, \eta) = 0$. The function f, thereby uniquely defined near x by the condition $f(\zeta) = \eta$, is continuous.

Proof. Let $T = d_2G(x, y)$ and, for $\alpha \in U$, $\beta \in V$, set

$$L(\alpha, \beta) = \beta - T^{-1}\big[G(\alpha, \beta)\big].$$

Of course T is invertible by hypothesis.

It follows by inspection that L is a continuous mapping from $U \times V$ to Z such that

$$L(x, y) = y - T^{-1}\big[G(x, y)\big] = y - T^{-1}(0) = y.$$

Also L is continuously differentiable in the second entry and

$$d_2L(x, y) = \mathrm{id} - T^{-1} \circ \big[d_2G(x, y)\big] = 0.$$

Since $d_2L(\cdot, \cdot)$ is a continuous function of its arguments, there is a product of balls $M \times N$ (with $M = B(x, r)$ and $N = B(y, s)$) about (x, y) on which $\|d_2L(\cdot, \cdot)\|$ is bounded by $1/2$. We may also assume, shrinking the ball M if necessary, that $\|L(\alpha, y) - y\| < s/2$ for $\alpha \in M$.

The mean-value theorem applied to L in its second variable (parametrize with a segment) now implies that $L(\alpha, \cdot)$ is a contraction, with constant $1/2$ (this statement is true uniformly in $\alpha \in M$). By Theorem 3.4.6, we conclude that for each $\zeta \in M$ there is a unique $\eta \in N$ such that $L(\zeta, \eta) = \eta$; moreover, the mapping $\zeta \mapsto \eta$ is continuous. Since $L(\zeta, \eta) = \eta$ if and only if $G(\zeta, \eta) = 0$, the result is proved. □

Remark 3.4.11 The use of the contraction mapping principle in proving the implicit function theorem is an intellectual descendent of the iterative proof first given in Goursat [Go 03]. Edouard Goursat (1858–1936) was inspired in turn by Charles Emile Picard's (1856–1941) iterative proof of the existence of solutions to ordinary differential equations.

3.5 The Rank Theorem and the Decomposition Theorem

The rank theorem is a variant of the implicit function theorem that is tailored to situations in which the Jacobian matrix of the mapping under study is of constant rank but not full rank. Consider, for instance, the example $F : \mathbf{R}^3 \to \mathbf{R}^3$ given by

$$F(x, y, z) = (x^3 + y + z, y^3 + y, x^3 - y^3 + z).$$

Then the Jacobian of F is

$$A = \begin{pmatrix} 3x^2 & 1 & 1 \\ 0 & 3y^2+1 & 0 \\ 3x^2 & -3y^2 & 1. \end{pmatrix}$$

Since the first and third columns of A are dependent, the rank of A is everywhere less than or equal to 2. On the other hand, the first and third rows of A are independent. Thus A has rank 2 at every point. If $b \in \mathbf{R}^3$ is a point in the range of F—for specificity, let us say that $b = (30, 10, 20)$—then we see that $F^{-1}(b)$ is a 1-dimensional set. For example, the point $(3, 2, 1)$ lies in $F^{-1}(b)$ and is an element of the variety

$$F^{-1}(b) = \{(x, 2, z) : z = -x^3 + 28\}.$$

This is a nonsingular curve of dimension 1. The geometric setup for other level sets is similar: In effect, the level sets of the mapping F foliate the domain of F.

Obversely, if we look at the image of F, then we see that it is a smooth 2-dimensional manifold given by $\{(x, y, z) : z = x - y\}$.

The rank theorem formalizes the observations that we have made for this specific mapping F in the context of a class of mappings having constant rank.

Theorem 3.5.1 (The Rank Theorem). *Let r, p, q be nonnegative integers and let $M = \mathbf{R}^{r+p}$, $N = \mathbf{R}^{r+q}$. Let $W \subseteq M$ be an open set and suppose that $F : W \to N$ is a continuously differentiable mapping. Assume that DF has rank r at each point of W.*

Fix a point $\mathbf{w} \in W$. There exist vector subspaces $M_1, M_2 \subseteq M$ and $N_1, N_2 \subseteq N$ such that $M = M_1 + M_2$, $N = N_1 + N_2$, $\dim M_1 = \dim N_1 = r$, each $\mathbf{m} \in M$ has a unique representation $\mathbf{m} = \mathbf{m}_1 + \mathbf{m}_2$ with $\mathbf{m}_j \in M_j$, each $\mathbf{n} \in N$ has a unique representation $\mathbf{n} = \mathbf{n}_1 + \mathbf{n}_2$ with $\mathbf{n}_j \in N_j$, and with the following properties: Set $F(\mathbf{x}) = F_1(\mathbf{x}) + F_2(\mathbf{x})$ with $F_1(\mathbf{x}) \in N_1$ and $F_2(\mathbf{x}) \in N_2$ for each $\mathbf{x} \in W$. Then there is an open set $U \subseteq W$ with $\mathbf{w} \in U$ such that

(1) *$F_1(U)$ is an open set in N_1;*

(2) *For each $\mathbf{n}_1 \in F_1(U)$ there is precisely one $\mathbf{n}_2 \in N_2$ such that*

$$\mathbf{n}_1 + \mathbf{n}_2 \in F(U).$$

We see that the rank theorem says in a very precise sense that the image $F(U)$ of F is a graph over $F_1(U)$, and can thereby be seen to be a smooth, r-dimensional surface (or manifold). The corresponding statements about the dimension and form of the level sets of F follow from a bit of further analysis, and we shall save those until after our consideration of the theorem.

As with many theorems in this subject, the continuously differentiable version is modeled on a rather transparent paradigm that comes from linear algebra. We first treat that special instance:

Lemma 3.5.2 *Let p, q, r be nonnegative integers and suppose that M, N are vector spaces with dimensions $r + p$ and $r + q$, respectively. Let $A : M \to N$ be a linear transformation, and assume that the rank of A is r. Then there exist vector spaces $M_1, M_2 \subseteq M$ and $N_1, N_2 \subseteq N$ such that*

(1) *Each $\mathbf{m} \in M$ can be written in a unique manner as $\mathbf{m} = \mathbf{m}_1 + \mathbf{m}_2$ with $\mathbf{m}_j \in M_j$, $j = 1, 2$;*

(2) *Each $\mathbf{n} \in N$ can be written in a unique manner as $\mathbf{n} = \mathbf{n}_1 + \mathbf{n}_2$ with $\mathbf{n}_j \in N_j$, $j = 1, 2$;*

(3) *$A\mathbf{m}_2 = 0$ for each $\mathbf{m}_2 \in M_2$;*

(4) *A maps M_1 to N_1 in a one-to-one, onto manner;*

(5) $\dim M_1 = \dim N_1 = r$.

Proof. The proof is just elementary linear algebra. Let N_1 be the image of A. Select a basis $\{\mathbf{v}_1, \ldots, \mathbf{v}_{r+q}\}$ for N such that $\{\mathbf{v}_1, \ldots, \mathbf{v}_r\}$ is a basis for N_1 (whose dimension we know to be r by the hypothesis on the rank of A). Define N_2 to be the span of $\{\mathbf{v}_{r+1}, \ldots, \mathbf{v}_{r+q}\}$.

For $j = 1, \ldots, r$, choose vectors $\mathbf{u}_j \in M$ such that $A\mathbf{u}_j = \mathbf{v}_j$. Then of course $\{\mathbf{u}_1, \ldots, \mathbf{u}_r\}$ is a linearly independent set, and we let M_1 be the span of that set. Let M_2 be the kernel of the operator A.

By fiat, property (2) now holds. Property (3) is true by definition, and so is property (4). Property (5) is equally obvious.

If $\mathbf{m} \in M$, then there are scalars $a_1, \ldots, a_r$ such that $A\mathbf{m} = \sum_{j=1}^{r} a_j \mathbf{v}_j$. Set $\mathbf{m}_1 = \sum_{j=1}^{r} a_j \mathbf{u}_j$ and $\mathbf{m}_2 = \mathbf{m} - \mathbf{m}_1$. Then $\mathbf{m}_1 \in M$ and $A\mathbf{m}_1 = A\mathbf{m}$. Therefore certainly $A\mathbf{m}_2 = 0$ so that $\mathbf{m}_2 \in M_2$.

Now properties (3) and (4) imply that $M_1 \cap M_2 = \emptyset$. Thus the representation $\mathbf{m} = \mathbf{m}_1 + \mathbf{m}_2$ is unique, so that (1) is proved. □

Now we will transfer this basic linear algebra fact to the continuously differentiable context.

Proof of Theorem 3.5.1. With $A = DF(\mathbf{w})$, $M = \mathbf{R}^{r+p}$, $N = \mathbf{R}^{r+q}$, let the spaces M_1, M_2, N_1, N_2 be as in the lemma. Let $T = A\big|_{M_1}$. By part (4) of the lemma, T is a linear isomorphism of M_1 onto N_1.

Since we will use the projections onto all four of M_1, M_2, N_1, and N_2 we introduce some notation for these mappings. Let P_i be the linear projection of M onto M_i given by $P_i(\mathbf{m}_1 + \mathbf{m}_2) = \mathbf{m}_i$ and let Q_j be the linear projection of N onto N_j given by $Q_j(\mathbf{n}_1 + \mathbf{n}_2) = \mathbf{n}_j$. With this notation, we have $F_j = Q_j F$, $A = AP_1 = Q_1 A = Q_1 A P_1$, $0 = AP_2 = Q_2 A = Q_2 A P_2$, $T^{-1}A = P_1$, and $AT^{-1}Q_1 = Q_1$.

Set

$$G(\mathbf{x}) = T^{-1}F_1(\mathbf{x}) + P_2\mathbf{x} = T^{-1}Q_1F(\mathbf{x}) + P_2\mathbf{x}, \qquad \mathbf{x} \in W. \tag{3.38}$$

Differentiating, we find

$$\begin{aligned} DG(\mathbf{w}) &= T^{-1}Q_1(DF(\mathbf{w})) + P_2 \\ &= T^{-1}(Q_1 A) + P_2 = T^{-1}A + P_2 = P_1 + P_2 . \end{aligned}$$

We conclude that $DG(\mathbf{w})$ is the identity mapping on M. We may now apply the inverse function theorem to the function G at the point $\mathbf{w}$. We conclude that there is a neighborhood U of $\mathbf{w}$ in W and a neighborhood V of $G(\mathbf{w})$ such that G is a one-to-one, onto, continuously differentiable mapping of U onto V. Taking subsets if necessary, we may assume that V is convex (this hypothesis is only for convenience, and is certainly not necessary).

Set $H = (G|_U)^{-1}$ and define

$$\Phi(\mathbf{z}) = F(H(\mathbf{z})) , \qquad \mathbf{z} \in V . \tag{3.39}$$

Since $AP_2 = 0$, (3.38) shows that $AG = AT^{-1}Q_1 F = Q_1 F$ so that

$$Q_1 F(H(\mathbf{z})) = AG(H(\mathbf{z})) = A\mathbf{z} = Q_1 A P_1 \mathbf{z} .$$

Therefore

$$\Phi(\mathbf{z}) \equiv Q_1 A P_1 \mathbf{z} + \phi(\mathbf{z}) \qquad \mathbf{z} \in V , \tag{3.40}$$

where $\phi(\mathbf{z}) \in N_2$.

By (3.39) and (3.40), $Q_1 F(U)$ is the set of all points of the form $Q_1 A P_1 \mathbf{z}$, $\mathbf{z} \in V$. Since V is open and N_1 is the range of A, part (1) of the theorem is proved.

To prove part (2) of the theorem, we need to show that $\Phi(\mathbf{z})$ depends only on $P_1\mathbf{z}$. So fix an element $\mathbf{z} \in V$. By (3.39) and (3.40), we know that

$$D\Phi(\mathbf{z}) = DF(H(\mathbf{z})) \cdot DH(\mathbf{z}) = Q_1 A P_1 + D\phi(\mathbf{z}) . \tag{3.41}$$

Since $DH(\mathbf{z})$ is an invertible linear operator on M, and since $DF(H(\mathbf{z}))$ has rank r, we see that the range R of $D\Phi(\mathbf{z})$ is a vector space of dimension r. Since the range of $D\phi(\mathbf{z})$ is in N_2, (3.41) shows that $Q_1 A P_1 = Q_1(D\Phi(\mathbf{z}))$. Hence Q_1 maps R into N_1; since both of the spaces R and N_1 have dimension r we conclude that $Q_1 : R \to N_1$ is also one-to-one and onto. Thus, for $\mathbf{z} \in V$,

$$Q_1 A P_1 \mathbf{h} = 0 \quad \text{implies} \quad D\Phi(\mathbf{z})\mathbf{h} = 0 . \tag{3.42}$$

If we now have the setup $\mathbf{z} \in V$, $\mathbf{h}_2 \in M_2$, $\mathbf{z} + \mathbf{h}_2 \in V$, then we define

$$\Lambda(t) = \Phi(\mathbf{z} + t\mathbf{h}_2) , \qquad 0 \le t \le 1 . \tag{3.43}$$

Because V is convex, this definition is valid and sensible. Since $A\mathbf{h}_2 = 0$, (3.42) and (3.43) imply that

$$D\Lambda(t) = D\Phi(\mathbf{z} + t\mathbf{h}_2)\mathbf{h}_2 = 0 , \qquad 0 \le t \le 1 .$$

Thus $\Lambda(1) = \Lambda(0)$ or $\Phi(\mathbf{z} + \mathbf{h}_2) = \Phi(\mathbf{z})$; this is what we needed to prove to establish (2). □

It is worth noting that if $\mathbf{n} \in N$ is a value of the mapping F and if $\mathbf{m} \in M$ is an element of $F^{-1}(\mathbf{n})$, then the Jacobian of $K(\mathbf{x}) \equiv F(\mathbf{x}) - \mathbf{n}$ (in the entries $\mathbf{v}_1, \ldots, \mathbf{v}_r$) with respect to the variables $\mathbf{u}_1, \ldots, \mathbf{u}_r$ has maximal rank. Therefore the implicit function theorem applies and we see that we can describe the level set as a smooth, parameterized p-dimensional manifold. This is often the practical significance of the rank theorem (although it is generally not formulated in this way).

It is not difficult to see that Theorem 3.5.1 is also true when the domain $\mathbf{R}^{r+p}$ and the range $\mathbf{R}^{r+q}$ of the mapping F are replaced by smooth manifolds. We leave the details to the reader (on a local coordinate patch, simply map the domain and range to appropriate Euclidean spaces).

We next formulate and prove a version of the inverse function theorem which says in effect that a given C^1 invertible mapping can be factored into elementary submappings. We begin with a definition. In what follows, let $\{\mathbf{e}_1, \ldots, \mathbf{e}_N\}$ be the canonical orthonormal basis for Euclidean space.

Definition 3.5.3 Let $E \subseteq \mathbf{R}^N$ be open and $F : E \to \mathbf{R}^N$ a mapping. Assume that, for some fixed, positive integer j,

$$\mathbf{e}_i \cdot F(\mathbf{x}) = \mathbf{e}_i \cdot \mathbf{x}$$

for all $\mathbf{x} \in E$ and $i \neq j$. This hypothesis simply says that $\mathbf{x}$ and $F(\mathbf{x})$ have the same i^{th} coordinate when $i \neq j$. In other words, F acts only in the jth coordinate. In the terminology of Rudin [Ru 64], such a mapping is called *primitive*.

A primitive mapping is a rather specialized object; in fact it appears to be too particular to bear much scrutiny. But the decomposition theorem that we now present tells us that any C^1 mapping may be factored into primitive mappings.

Theorem 3.5.4 *Suppose that F is a C^1 mapping of an open set $E \subseteq \mathbf{R}^N$ into $\mathbf{R}^N$. Assume that $0 \in E$, $F(0) = 0$, and det $DF(0) \neq 0$. Then there is a neighborhood U of 0 in $\mathbf{R}^N$ in which the representation*

$$F(\mathbf{x}) = G_N(B_N(G_{N-1}(B_{N-1} \cdots G_1(B_1(\mathbf{x}) \cdots))))$$

is valid. Here each G_j is a primitive C^1 mapping on U, $G_j(0) = 0$, and each B_j is a linear operator on $\mathbf{R}^N$ which is either the identity or which interchanges two coördinates.

Proof. The proof proceeds by building on the number of mappings in the factorization. We will construct a sequence of mappings F_m which come closer and closer to satisfying the conclusion of the theorem. The inductive statement $(\mathcal{P}_m)$ will be

- *The mapping F_m maps a neighborhood U_m of $0 \in \mathbf{R}^N$ into $\mathbf{R}^N$, $F_m(0) = 0$, F_m is of class C^1, $A_m \equiv DF_m(0)$ is invertible and*

$$\mathbf{e}_i \cdot F_m(\xi) = \mathbf{e}_i \cdot \xi \tag{3.44}$$

holds for $\xi \in U_m$ and $1 \leq i < m$.

For $m = 1$ we set $F_1 = F$ and note that $(\mathcal{P}_1)$ is obviously true since there are then no $\mathbf{e}_i$ with $1 \leq i < m$.

Assume now that the $(\mathcal{P}_\nu)$ has been proved for $1 \leq \nu \leq m$. We now establish $(\mathcal{P}_{m+1})$. Set $\alpha_{ij} = \mathbf{e}_i \cdot A_m \mathbf{e}_j$. Then (3.44) tells us that $\alpha_{ij} = 0$ if $i < m \leq j$. If we also knew that $\alpha_{mj} = 0$ for all $j \geq m$, then the representation $A_m \mathbf{e}_j = \sum_i \alpha_{ij} \mathbf{e}_i$ would show that the collection $A_m \mathbf{e}_m, \ldots, A_m \mathbf{e}_N$ of $N+1-m$ linearly independent vectors lies in the span of the $N - m$ vectors $\mathbf{e}_{m+1}, \ldots \mathbf{e}_N$; that is a contradiction. Thus there is an index j, $m \leq j \leq N$, such that $\alpha_{mj} \neq 0$. Fix this j.

Define projection operators P_m such that $P_m \mathbf{e}_i = \mathbf{e}_i$ if $i \neq m$ and $P_m \mathbf{e}_m = 0$. Also define linear operators B_m satisfying $B_m \mathbf{e}_m = \mathbf{e}_j$, $B_m \mathbf{e}_j = \mathbf{e}_m$, and $B_m \mathbf{e}_i = \mathbf{e}_i$ for $i \neq j, i \neq m$. Put

$$G_m(\mathbf{x}) = P_m \mathbf{x} + [\mathbf{e}_m \cdot F_m(B_m(\mathbf{x}))]\mathbf{e}_m \,. \tag{3.45}$$

Then G_m is obviously primitive. Since $D(F_m B_m)(0) = A_m B_m$, we see by differentiation that

$$DG_m(0)\mathbf{h} = P_m \mathbf{h} + [\mathbf{e}_m \cdot A_m B_m \mathbf{h}]\mathbf{e}_m \quad \text{for } \mathbf{h} \in \mathbf{R}^N \,.$$

If $DG_m(0)\mathbf{h} = \mathbf{0}$, then the last line shows that $P_m \mathbf{h} = \mathbf{0}$, so that $\mathbf{h} = \lambda \mathbf{e}_m$ for some scalar λ. But we also know that $\mathbf{e}_m \cdot A_m B_m \mathbf{h} = 0$, or $\lambda \alpha_{mj} = 0$ by the definition of B_m. Since $\alpha_{mj} \neq 0$, we see that $\lambda = 0$, hence $\mathbf{h} = \mathbf{0}$.

We thus see that $DG_m(0)$ is one-to-one. Therefore it is invertible, and the implicit function theorem then implies that G_m is one-to-one on a neighborhood U_m of 0. Also $G_m(U_m) \equiv V_m$ is an open subset of $\mathbf{R}^N$. Define

$$F_{m+1}(\mathbf{y}) = F_m(B_m G_M^{-1}(\mathbf{y})) \,, \qquad \mathbf{y} \in V_m \,. \tag{3.46}$$

If $\mathbf{y} \in V_m$ with $\mathbf{y} = G_m(\mathbf{x})$, $\mathbf{x} \in U_m$, then (3.45) shows that

$$\mathbf{e}_m \cdot \mathbf{y} = \mathbf{e}_m \cdot F_m(B_m \mathbf{x}) \qquad \text{and} \qquad \mathbf{e}_i \cdot \mathbf{y} = \mathbf{e}_i \cdot \mathbf{x} \qquad \text{for } i < m \,. \tag{3.47}$$

As a result, (3.44) and the definition of B_m imply that

$$\mathbf{e}_i \cdot F_{m+1}(\mathbf{y}) = \mathbf{e}_i \cdot B_m \mathbf{x} = \mathbf{e}_\mathbf{x} \cdot \mathbf{x} = \mathbf{e}_i \cdot \mathbf{y}$$

if $i < m$. Also (3.47) implies that

$$\mathbf{e}_m \cdot F_{m+1}(\mathbf{y}) = \mathbf{e}_m \cdot F_m(B_m \mathbf{x}) = \mathbf{e}_m \cdot \mathbf{y} \,.$$

We see as a consequence of these calculations that F_{m+1} satisfies $(\mathcal{P}_{m+1})$. Rewriting (3.46) in the form

$$F_m(\mathbf{x}) = F_{m+1}(G_m(B_m\mathbf{x}))\,, \qquad m = 1, 2, \ldots, N,$$

and noting that F_{N+1} is the identity, we finally see that

$$\begin{aligned} F(\mathbf{x}) = F_1(\mathbf{x}) &= F_2[G_1(B_1\mathbf{x})] \\ &= F_3[G_2(B_2G_1(B_1\mathbf{x}))] \\ &\vdots \\ &= F_{N+1}[G_N(B_N(G_{N-1}(B_{N-1}\cdots G_1(B_1(\mathbf{x})\cdots)))))] \\ &= G_N(B_N(G_{N-1}(B_{N-1}\cdots G_1(B_1(\mathbf{x})\cdots))))\,, \end{aligned}$$

which gives the conclusion we desire. □

3.6 A Counterexample

Example 3.6.1 *This example will show that one cannot omit the hypothesis in the inverse function theorem requiring the derivative to be continuous.* Consider $f(x) = \alpha x + x^2 \sin(1/x)$, where $0 < \alpha < 1$. Extend f to $x = 0$ by setting $f(0) = 0$. This function is differentiable everywhere and $f'(0) = \alpha \neq 0$. In fact, we compute

$$f'(x) = \alpha + 2x\sin(1/x) - \cos(1/x) \text{ for } x \neq 0, \tag{3.48}$$

while the derivative at 0 is obtained by directly examining the limit of the difference quotient.

We note that f' is *not* continuous at zero, so the hypothesis in the inverse function theorem that the function be C^1 is *not* satisfied. We will show below that the f does not have an inverse function in any neighborhood of 0.

It is clear from freshman calculus that at any point where $f'(x) = 0$ and $f''(x) \neq 0$, there cannot be a local inverse. We claim that there are infinitely many such points in any neighborhood of 0. From (3.48) and the fact that $|\alpha| < 1$, it is clear that there are infinitely many zeros of f' in any neighborhood of 0. It remains to show that such zeros of f' are not also zeros of f''. We will prove this assertion by contradiction.

We compute

$$f''(x) = (2 - 1/x^2)\sin(1/x) - (2/x)\cos(1/x) \text{ for } x \neq 0. \tag{3.49}$$

If *both* $f'(x) = 0$ and $f''(x) = 0$ hold for some $x \neq 0$, then the system of simultaneous linear equations

$$\begin{aligned} 2xS - C &= -\alpha \\ (2 - 1/x^2)S - (2/x)C &= 0 \end{aligned}$$

has the solution $S = \sin(1/x)$, $C = \cos(1/x)$. On the other hand, we can apply Cramer's rule to the linear system to conclude that

$$S = \alpha \frac{-2x}{1+2x^2}, \tag{3.50}$$

$$C = \alpha \frac{1-2x^2}{1+2x^2}. \tag{3.51}$$

If we take S and C to be given by (3.50) and (3.51), then we see that

$$S^2 + C^2 = \alpha^2 \frac{1+4x^4}{(1+2x^2)^2} \tag{3.52}$$

and, for small nonzero values of x, the right-hand side of (3.52) is not equal to 1. Thus S and C cannot be equal to $\sin(1/x)$ and $\cos(1/x)$, respectively, and, consequently, x cannot be a zero both of f' and of f''. □

4 Applications

4.1 Ordinary Differential Equations

There is a strong connection between the implicit function theorem and the theory of differential equations. This is true even from the historical point of view, for Picard's iterative proof of the existence theorem for ordinary differential equations inspired Goursat to give an iterative proof of the implicit function theorem (see Goursat [Go 03]). In the mid-twentieth century, John Nash pioneered the use of a sophisticated form of the implicit function theorem in the study of partial differential equations. We will discuss Nash's work in Section 6.4. In this section, we limit our attention to ordinary (rather than partial) differential equations because the technical details are then so much simpler. Our plan is first to show how a theorem on the existence of solutions to ordinary differential equations can be used to prove the implicit function theorem. Then we will go the other way by using a form of the implicit function theorem to prove an existence theorem for differential equations.

A typical existence theorem for ordinary differential equations is the following fundamental result[1] (see for example, Hurewicz [Hu 64]):

Theorem 4.1.1 (Picard) *If $F(t,x)$, $(t,x) \in \mathbf{R} \times \mathbf{R}^N$, is continuous in the $(N+1)$-dimensional region $(t_0 - a, t_0 + a) \times \mathbf{B}(x_0, r)$, then there exists a solution $x(t)$*

[1] This fundamental theorem is commonly known as *Picard's existence and uniqueness theorem*. The classical proof uses a method that has come to be known as the *Picard iteration technique*. See [Pi 93].

of

$$\frac{dx}{dt} = F(t, x), \quad x(t_0) = x_0, \tag{4.1}$$

defined over an interval $(t_0 - h, t_0 + h)$.

Remark 4.1.2 The solution of (4.1) need not be unique if F is only continuous. For example, the problem of finding $x(t)$ satisfying $x' = x^{2/3}$, $x(0) = 0$, has the two solutions $x \equiv 0$ and $x(t) = (t/3)^3$. To guarantee that the solution of (4.1) is unique, it is sufficient to assume additionally that F satisfies a Lipschitz condition as a function of x

We can give an alternative proof of the implicit function theorem as a corollary of Theorem 4.1.1.

Theorem 4.1.3 *Suppose that* $U \subset \mathbf{R}^{N+1}$ *and that* $H : U \to \mathbf{R}$ *is* C^1. *If* $H(t_0, x_0) = 0$, $(t_0, x_0) \in \mathbf{R} \times \mathbf{R}^N$, *and the* $N \times N$ *matrix*

$$\left(\frac{\partial H_i}{\partial x_j}(t_0, x_0)\right)_{i,j=1,2,\ldots,N}$$

is nonsingular, then there exists an open interval (t_0-h, t_0+h) *and a continuously differentiable function* $\phi : (t_0 - h, t_0 + h) \to \mathbf{R}^N$ *such that* $\phi(t_0) = x_0$ *and*

$$H(t, \phi(t)) = 0.$$

Proof. We consider the case $N = 1$ in some detail. First, choose $a, r > 0$ so that $(t_0 - a, t_0 + a) \times (x_0 - r, x_0 + r) \subseteq U$ and $(\partial H/\partial x)(t, x)$ is nonvanishing on $(t_0-a, t_0+a) \times (x_0-r, x_0+r)$. Then define $F : (t_0-a, t_0+a) \times (x_0-r, x_0+r) \to \mathbf{R}$ by setting

$$F(t, x) = -\frac{\partial H}{\partial t}(t, x) \Big/ \frac{\partial H}{\partial x}(t, x). \tag{4.2}$$

Since F is continuous, we can apply Theorem 4.1.1 to conclude that there exists a solution of the problem

$$\frac{dx}{dt} = F(t, x), \quad x(t_0) = x_0$$

defined on an interval $(t_0 - h, t_0 + h)$. We define $\phi : (t_0 - h, t_0 + h) \to \mathbf{R}$ by setting

$$\phi(t) = x(t).$$

Note that

$$\phi(t_0) = x_0 \tag{4.3}$$

and

$$\begin{aligned}\phi'(t) &= \frac{dx}{dt}(t) = F(t, x(t)) \\ &= -\frac{\partial H}{\partial t}(t, \phi(t)) \Big/ \frac{\partial H}{\partial x}(t, \phi(t)).\end{aligned} \tag{4.4}$$

By (4.3), we have $H(t_0, \phi(t_0)) = H(t_0, x_0) = 0$, and by (4.4), we have

$$\frac{d}{dt} H(t, \phi(t)) = \frac{\partial H}{\partial t}(t, \phi(t)) + \phi'(t) \frac{\partial H}{\partial x}(t, \phi(t)) = 0.$$

Thus we have $H(t, \phi(t)) = 0$ on the interval $(t_0 - h, t_0 + h)$.

In case $N > 1$, we choose $a, r > 0$ so that $(t_0 - a, t_0 + a) \times \mathbb{B}(x_0, r) \subseteq U$ and so that the $N \times N$ matrix

$$D_x H = \left(\frac{\partial H_i}{\partial x_j} \right)_{i,j=1,2,\ldots,N}$$

is nonsingular on $(t_0 - a, t_0 + a) \times \mathbb{B}(x_0, r)$. Next, we replace (4.2) by

$$F(t, x) = -\left[D_x H(t + t_0, x + x_0) \right]^{-1} \left(\frac{\partial H}{\partial t}(t + t_0, x + x_0) \right).$$

The proof then proceeds as before. □

Remark 4.1.4 The proof of Theorem 4.1.3 given above is clearly limited to the case of one independent variable in the implicitly defined function. The case of one dependent variable and several independent variables can be obtained by replacing (4.2) with the appropriate system of first-order partial differential equations. The system of partial differential equations is solved by applying the existence theorem for ordinary differential equations (Theorem 4.1.1), *with parameters*, to each independent variable in turn. For example, if we have the equation $H(x, y, z) = 0$ which we are considering near a point (x_0, y_0, z_0) where H is zero and $\partial H / \partial z$ is nonzero (so the implicit function $z(x, y)$ will involve two independent variables), then there will be two first-order partial differential equations that $z(x, y)$ must satisfy:

$$\frac{\partial z}{\partial x} = -\frac{\partial H}{\partial x} \Big/ \frac{\partial H}{\partial z}, \tag{4.5}$$

$$\frac{\partial z}{\partial y} = -\frac{\partial H}{\partial y} \Big/ \frac{\partial H}{\partial z}. \tag{4.6}$$

Restricting to a neighborhood of (x_0, y_0, z_0) in which $\partial H / \partial z$ is non-vanishing will enable us to conclude by an appeal to Rolle's theorem that the function $z(x, y)$ is uniquely defined, without the hypotheses for the uniqueness of solutions of ordinary differential equations.

The *second* equation, (4.6), is solved by solving the *first* equation, (4.5), with $y = y_0$ fixed and with the initial condition $z = z_0$. Then the resulting function $z(x, y_0)$ is used to provide the initial condition for (4.6); the initial value problem is then solved while treating x as a parameter. This process will produce a solution of (4.6) in an open set about the point (x_0, y_0). By carrying out the same process, but in the other order, we can obtain a solution of (4.5) in an open set about the same point (x_0, y_0). Because of the form of the right-hand sides of (4.5) and (4.6), those two solutions will be consistent and will define just one function that satisfies *both* equations.

Finally, the general implicit function theorem for any number of dependent and independent variables can be proved using Dini's induction procedure (see Section 3.2). □

We have seen that the implicit function theorem can be treated, in a sense, as a corollary of the existence theorem for ordinary differential equations. What we would like to do next is prove the converse: that we can use the implicit function theorem to prove the existence of solutions to ordinary differential equations. The Banach space methods of Section 3.4 will be required for this argument. We recall the statement of the theorem:

Theorem 4.1.1 *If $F(t, x)$, $(t, x) \in \mathbf{R} \times \mathbf{R}^N$, is continuous in the $(N + 1)$-dimensional region $(t_0 - a, t_0 + a) \times \mathbf{B}(x_0, r)$, then there exists a solution $x(t)$ of*

$$\frac{dx}{dt} = F(t, x), \quad x(t_0) = x_0, \tag{4.1}$$

defined over an interval $(t_0 - h, t_0 + h)$.

Proof. For convenience of notation, let us suppose that $t_0 = 0$.

Let $\mathcal{B}_0$ be the space of bounded continuous $\mathbf{R}^N$-valued functions on $(-a, a)$, normed by the supremum of the magnitude of the function. Let $\mathcal{B}_1$ be the space of bounded continuously differentiable $\mathbf{R}^N$-valued functions on $(-a, a)$ that also have a bounded derivative. We norm this space by the sum of the supremum of the magnitude of the function and the supremum of the magnitude of the derivative of the function. We define a map $\mathcal{F} : \mathcal{B}_1 \to \mathcal{B}_0 \times \mathbf{R}$ by setting

$$\mathcal{F}\Big[x(t)\Big] = \Big[x'(t) - F(t, x(t)),\ x(0) - x_0\Big].$$

With this notation, a solution of (4.1) is given by a zero of $\mathcal{F}$.

We imbed the problem of solving $\mathcal{F}[x] = [0, 0]$ into a larger problem Define $\mathcal{H} : [0, 1] \times \mathcal{B}_1 \to \mathcal{B}_0 \times \mathbf{R}$ by setting

$$\mathcal{H}\Big[\alpha,\ X(\tau)\Big] = \Big[X'(\tau) - \alpha F(\alpha\tau, X(\tau)),\ X(0) - x_0\Big].$$

We observe that

$$\mathcal{H}[0,\ x_0] = [0,\ 0], \tag{4.7}$$

where x_0 in (4.7) represents the constant function. Also, we observe that the Fréchet derivative of $\mathcal{H}$ at $[0, x_0]$ is given by $X \mapsto X'$. It follows from the implicit function theorem for Banach spaces, Theorem 3.4.10, that for all small enough choices of α there exists an $X(\alpha, \tau)$ such that

$$D_\tau X(\alpha, \tau) - \alpha F(\alpha\tau, X(\alpha, \tau)) = 0, \quad X(\alpha, 0) = x_0.$$

For such an $\alpha > 0$, we define $x(t)$ by setting

$$x(t) = X(\alpha, t/\alpha).$$

It follows that

$$x'(t) = \frac{1}{\alpha} \cdot D_\tau(\alpha, t/\alpha) = \frac{1}{\alpha} \cdot \alpha \cdot F[\alpha(t/\alpha), X(\alpha, t/\alpha)] = F(t, x(t)).$$

Thus our differential equation is solved, and the theorem is proved. □

4.2 Numerical Homotopy Methods

Suppose we wish to solve a system of nonlinear equations

$$F(x) = 0 \tag{4.8}$$

where $F : \mathbf{R}^N \to \mathbf{R}^N$ is smooth. Only in very special circumstances will it be possible to solve (4.8) in closed form; generally, numerical methods must be employed and an approximate solution thereby obtained. Of course, we would probably like to apply Newton's method, but for that we need a reasonable starting point for the iteration. In case we do not have such a reasonable starting point for Newton's method, some alternative procedure is needed. One such method is the *homotopy method* (also called the *continuation* or *imbedding* method).

In the homotopy method, we imbed the problem of interest, (4.8), into a larger problem of finding the zeros of a function $H : \mathbf{R}^{N+1} \to \mathbf{R}^N$. However, the function H is to be specially chosen so that the function $F_0 : \mathbf{R}^N \to \mathbf{R}^N$ defined by setting

$$F_0(x) = H(0, x) \tag{4.9}$$

is one that we understand well, while the function F in which we are interested is given by

$$F(x) = H(1, x).$$

The plan then is to follow the zeros of H from a starting point $(0, x_0) \in \mathbf{R}^{N+1}$ with $F_0(x_0) = 0$ along a curve $(t(s), x(s))$, $0 \leq s \leq 1$, for which

$$H(t(s), x(s)) = 0, \tag{4.10}$$

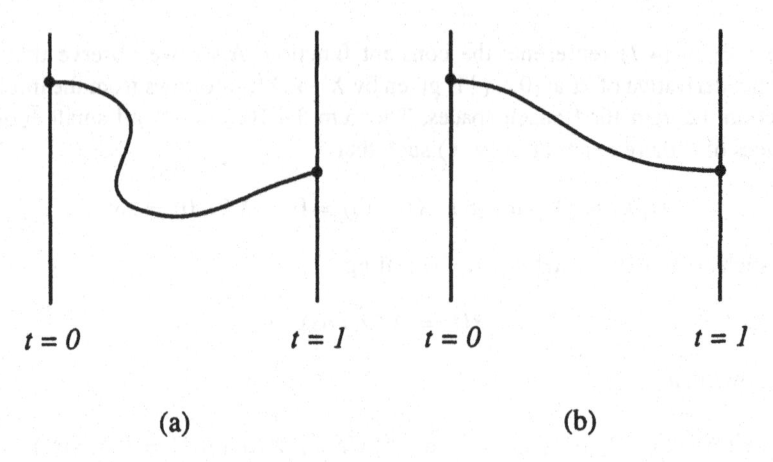

Figure 4.1. Nice Continuation

to a point $(1, x_1) = (t(1), x(1))$ where we will have $F(x_1) = 0$. This will solve the original problem (4.8). Two standard choices for H are the *convex homotopy* defined by setting

$$H(t, x) = (1 - t)F_0(x) + tF(x)$$

and the *global homotopy* defined by setting

$$H(t, x) = F(x) - (1 - t)F(x_0),$$

where x_0 can be any convenient value.

The picture we would like to see for the curve $(t(s), x(s))$ along which (4.10) holds should resemble that in Figure 4.1(a). It would be even better if the curve resembled that in Figure 4.1(b), because in that case we could parameterize the curve by t itself. On the other hand, it is conceivable that the solution set of

$$H(t, x) = 0$$

might look like that in Figure 4.2 where, starting from a zero of the form $H(0, x_0)$, we can never arrive at a zero of the form $H(1, x_1)$. Notice that there are four types of bad behavior for $\{(t, x) : H(t, x) = 0\}$ in Figure 4.2: (1) A curve starts at $t = 0$, but doubles back without ever getting to $t = 1$, (2) a curve becomes unbounded in x, (3) a curve reaches a bifurcation point where curves cross, and (4) a curve comes to a dead end where it cannot be continued. All of these instances of bad behavior are possible; nonetheless they all can be ruled out by imposing some simple hypotheses and applying the implicit function theorem.

To illustrate the ideas, we first state a theorem in which we can show that the curve $H(t(s), x(s)) = 0$ has the nice form shown in Figure 4.1(b).

Theorem 4.2.1 *Let U be an open subset of $\mathbf{R}^N$. Suppose that H is continuously differentiable in an open set containing $[0, 1] \times U$, that the function F_0 given by*

$$F_0(x) = H(0, x)$$

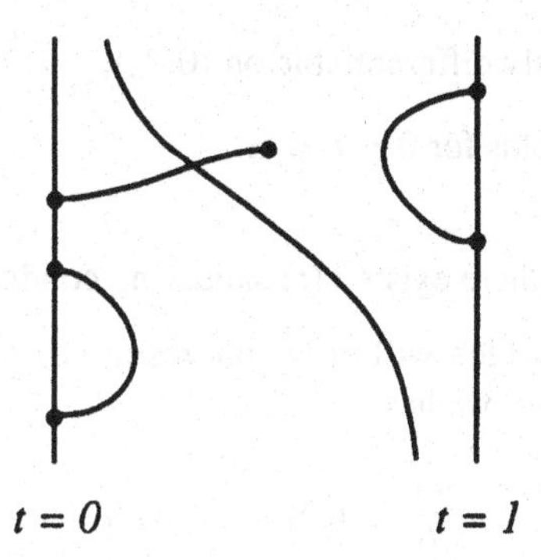

Figure 4.2. Bad Continuation

has a zero x_0 *in* U, *and that* $d > 0$ *is such that*

$$\{x : |x - x_0| < d\} \subseteq U.$$

If, on all of $[0, 1] \times U$, *the matrix*

$$D_x H = \left(\frac{\partial H_i}{\partial x_j}\right)_{i,j=1,2,\ldots,N}$$

is nonsingular and

$$\left|(D_x H)^{-1} D_t H\right| < d$$

holds, where $D_t H$ *is the column vector*

$$D_t H = \left(\frac{\partial H_i}{\partial t}\right)_{i=1,2,\ldots,N},$$

then there is a continuously differentiable function $x(t)$, *defined for* $0 \le t \le 1$, *satisfying*

$$H(t, x(t)) = 0. \tag{4.11}$$

In particular, we have

$$F(x(1)) = H(1, x(1)) = 0.$$

Proof. By the implicit function theorem applied at $(0, x_0)$, there exists a continuously differentiable function $x(t)$ defined in a neighborhood of 0 with $x(0) = x_0$ and satisfying (4.11) on that neighborhood. Thus there exists a positive number t_0 such that

(1) $x(t)$ is defined on the half open interval $[0, t_0)$,

(2) $x(0) = x_0$,

(3) $x(t)$ is continuously differentiable on $(0, t_0)$,

(4) $H(t, x(t)) = 0$ holds for $0 \le t < t_0$.

Set

$$t^* = \sup \{t_0 : \text{there exists } x(t) \text{ satisfying conditions (1)–(4)}\}.$$

We claim that $1 < t^*$, which will prove the result. To see this, note that, by the implicit function theorem, we have

$$\frac{dx}{dt} = (D_x H)^{-1} D_t H,$$

so that

$$|x'(t)| < d$$

holds whenever $t \le 1$ and $x'(t)$ exists and $x(t)$ lies in U. It follows that, if it were the case that $t^* \le 1$, then $x^* = \lim_{t \uparrow t^*} x(t)$ exists. Now we can apply the implicit function theorem again, but at the point (t^*, x^*), and thus extend $x(t)$ to a larger interval. This contradicts the definition of t^*. □

Example 4.2.2 Consider the equation

$$1 + x + e^x = 0. \tag{4.12}$$

If we define

$$H(t, x) = 1 + x + te^x,$$

then all the hypotheses of the theorem are satisfied by letting the interval $(-2, 0)$ play the role of the set U. Thus there exists a function $x(t)$, defined for $0 \le t \le 1$, with $x(0) = -1$ and

$$1 + x(t) + te^{x(t)} = 0. \tag{4.13}$$

In particular, $x(1) \approx -1.278465$ solves the equation (4.12). The curve is shown in Figure 4.3.

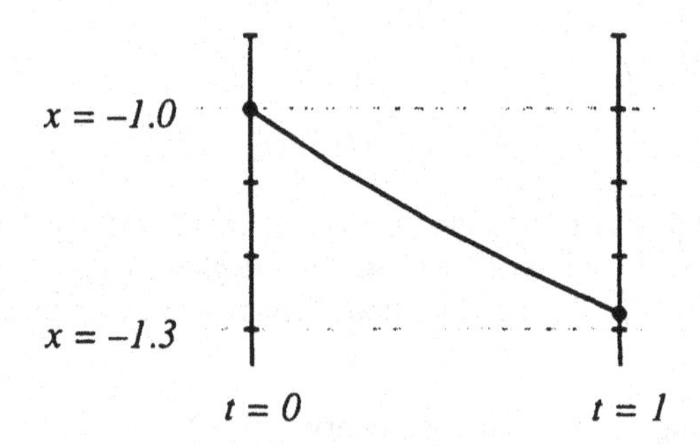

Figure 4.3. Continuation Curve

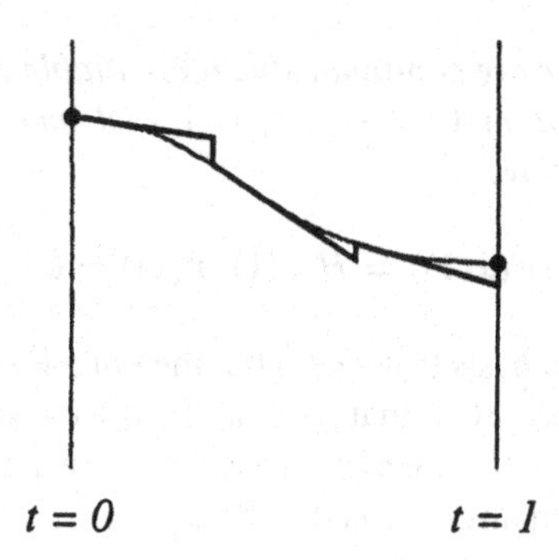

Figure 4.4. Predictor-Corrector Path

As was noted in the proof of Theorem 4.2.1, we know that $x(t)$ is the solution of the initial value problem

$$\frac{dx}{dt} = -\frac{e^x}{1+te^x}, \quad x(0) = -1. \tag{4.14}$$

The observation that the curve satisfies an initial value problem for an ordinary differential equation also can be used as a tool in finding $x(1)$ by a predictor–corrector method: Imagine we are incrementing t by a given step-size Δt. The differential equation (4.14) can be used to predict a reasonable approximation of $x(t + \Delta t)$ simply by using an "Euler step"

$$\Delta x = \frac{dx}{dt}\Delta t.$$

Then a correction can be made by using the equation (4.13). With $\Delta t = \frac{1}{3}$, this predictor–corrector process would produce a curve that is represented schematically by the jagged approximation to the smooth curve shown in Figure 4.4. In fact, for the particular example (4.13) and for $\Delta t = \frac{1}{3}$, the correction back to the curve in Figure 4.3 is so small as to be imperceptible. □

The hypotheses of the previous theorem rule out all possible bad behavior and insure that the curve $(t(s), x(s))$ never doubles back in the t direction, so it has the simple structure $x = x(t)$. A more general, but still nice, situation is provided by the next theorem.

Theorem 4.2.3 *Let U be a bounded open subset of $\mathbf{R}^N$. Suppose that H is continuously differentiable in an open set containing $[0, 1] \times \overline{U}$, that the function F_0 given by*

$$F_0(x) = H(0, x)$$

has a unique zero x_0 in U, and that $H(t, x) \neq 0$ holds for all $(t, x) \in [0, 1] \times \partial U$. If $DH(t, x)$ is of rank N for $(t, x) \in [0, 1] \times U$ and if the matrix

$$\left(\frac{\partial H_i}{\partial x_j}(0, x_0)\right)_{i,j=1,2,\ldots,N}$$

is nonsingular, then there are continuously differentiable functions $t(s)$ *and* $x(s)$, *defined for* $0 \le s \le 1$, *with* $t(0) = 0$, $t(1) = 1$, *and such that* $H(t(s), x(s)) = 0$ *holds. In particular, we have*

$$F(x(1)) = H(t(1), x(1)) = 0.$$

Remark 4.2.4 The hypothesis that $F_0(x)$ has the unique zero x_0 rules out the possibility that the curve $(t(s), x(s))$ that starts at $(0, x_0)$ and satisfies $H(t(s), x(s)) = 0$ could ever return to $\{0\} \times U$. The hypothesis that there are no zeros on $[0, 1] \times \partial U$ insures that the curve remains confined to $[0, 1] \times U$. Finally, the hypothesis that DH is of rank N guarantees that the set of points (t, x) for which $H(t, x) = 0$ really is a curve and that it cannot come to a bifurcation point or to a dead end. Thus the curve must emerge somewhere and, since all other possibilities are excluded, it must emerge at a point of the form $(1, x_1)$ with $x_1 \in U$.

Proof. Our main tool for constructing the curve will be Theorem 4.3.1 (to be treated later), which gives various equivalent definitions for a smooth surface or, in this case, a smooth curve. The proofs in Section 4.3 do not use the results of this section.

We will begin with an arc-length parametrization, and at the end we make a change of variables to obtain the parametrization required in the statement of the theorem.

By the hypothesis that DH is of rank N, we can apply Theorem 4.3.1 to conclude that, in a neighborhood of any point (t^*, x^*) with $H(t^*, x^*) = 0$, there exists a small section of curve satisfying

$$H(t(\sigma), x(\sigma)) = 0.$$

In particular, there is a section of curve through $(0, x_0)$. Let us suppose that this section of curve is parametrized by arc-length with $t(0) = 0$ and $x(0) = x_0$. Because

$$\left(\frac{\partial H_i}{\partial x_j}(0, x_0)\right)_{i,j=1,2,\ldots,N}$$

is nonsingular, we can conclude that $t'(0) \neq 0$. To see this, note that we have

$$(D_t H)\, t'(0) + (D_x H)\, x'(0) = 0,$$

so

$$x'(0) = -t'(0)\, (D_x H)^{-1}\, D_t H.$$

Thus, if if it were the case that $t'(0) = 0$, then the curve $(t(\sigma), x(\sigma))$ would have vanishing velocity, and that is not allowed under Theorem 4.3.1(1). Making a change of variable from σ to $-\sigma$ if necessary, we can suppose that $t'(0) > 0$.

By the preceding argument, we see that there exists a positive number σ_0 such that

(1) $t(\sigma)$ and $x(\sigma)$ are defined on the half open interval $[0, \sigma_0)$,

(2) $t(0) = 0$ and $x(0) = x_0$,

(3) $t(\sigma)$ and $x(\sigma)$ are continuously differentiable on $(0, \sigma_0)$,

(4) $(t(\sigma), x(\sigma))$ is an arc-length parametrization,

(5) $H(t(\sigma), x(\sigma)) = 0$ holds for $0 \le \sigma < \sigma_0$.

Set

$$\sigma^* = \sup\{\sigma_0 : \text{there exist } (t(\sigma), x(\sigma)) \text{ satisfying conditions (1)–(5)}\}.$$

We claim that $1 < \sup_{\sigma \in [0,\sigma^*)} t(\sigma)$. If not, we let (t^*, x^*) be an accumulation point of $(t(\sigma), x(\sigma))$ and apply Theorem 4.3.1 at (t^*, x^*). We conclude that, in a sufficiently small neighborhood of (t^*, x^*), the set of points (t, x) for which $H(t, x) = 0$ is a curve of finite length. Hence $\sigma^* < \infty$ and the curve $(t(\sigma), x(\sigma))$, $0 \le \sigma < \sigma^*$, can be extended past σ^*. This contradicts the definition of σ^*.

Finally, we set $\sigma_1 = \inf\{\sigma : t(\sigma) = 1\}$ and reparametrize by setting $s = \sigma/\sigma_1$. □

The preceding proof applies equally well with slightly weaker hypotheses, so we state that result as a corollary.

Corollary 4.2.5 *Under the circumstances of Theorem 4.2.3, if all the hypotheses of that theorem hold except that $DH(t, x)$ is assumed to be of rank N for all $(t, x) \in [0, 1] \times U \cap \{(t, x) : H(t, x) = 0\}$ rather than for all $(t, x) \in [0, 1] \times U$, then the conclusion of Theorem 4.2.3 still holds.*

When, in a particular case, we wish to apply Theorem 4.2.3, we might find that it is not true that DH is of rank N in the whole region $[0, 1] \times U$. It might not even be true that DH is of rank N on the set

$$\{(t, x) : H(t, x) = 0\}.$$

Nonetheless, if H is C^2 and we are willing to replace the value $0 \in \mathbf{R}^N$ by some nearby value, then we can arrange for DH to be of rank N on the inverse image of that value. This is a consequence of Sard's theorem. For completeness, we state Sard's theorem next.

Definition 4.2.6 Let $U \subseteq \mathbf{R}^M$, and $V \subseteq \mathbf{R}^K$ be open sets. If $H : U \to V$ is a C^k function, then we say $y \in V$ is a *critical value* for H if there exists $x \in U$ with $H(x) = y$ and for which the rank of the matrix

$$\left(\frac{\partial H_i}{\partial x_j}\right)_{i=1,2,\ldots,K,\,j=1,2,\ldots,M}$$

is less than K.

Theorem 4.2.7 (Sard) *Let $U \subseteq \mathbf{R}^M$, and $V \subseteq \mathbf{R}^K$ be open sets. If $H : U \to V$ is a C^k function, where $k \geq M/K$, then the set of critical values of H has Lebesgue K-dimensional measure zero.*

The reader can find a proof of Sard's theorem for the case in which H is C^∞ in Krantz and Parks [KP 99; Section 5.1]. A more general form of the theorem together with its proof can be found in Federer [Fe 69; Section 3.4].

In our present context of numerical homotopy methods, we have the following result.

Theorem 4.2.8 *Let U be a bounded open subset of $\mathbf{R}^N$. Suppose that H is C^2 in an open set containing $[0, 1] \times \overline{U}$, that the function F_0 given by*

$$F_0(x) = H(0, x)$$

has a unique zero x_0 in U, and that $H(t, x) \neq 0$ holds for all $(t, x) \in [0, 1] \times \partial U$. If the matrix

$$\left(\frac{\partial H_i}{\partial x_j}(0, x_0) \right)_{i,j=1,2,\ldots,N}$$

is nonsingular then, for almost every choice of c in a sufficiently small neighborhood of 0, there are continuously differentiable functions $t(s)$ and $x(s)$, defined for $0 \leq s \leq 1$, with $t(0) = 0$, $t(1) = 1$, and such that $H(t(s), x(s)) = c$ holds. In particular, we have

$$F(x(1)) = H(t(1), x(1)) = c.$$

Proof. We claim that, for all choices of c in a sufficiently small neighborhood of $0 \in \mathbf{R}^N$, there is a unique x in U with $F_0(x) = c$. Because $D_x F_0$ is nonsingular at x_0, there is a neighborhood V of x_0 that maps one-to-one and onto a neighborhood W of $0 \in \mathbf{R}^N$. So, for c sufficiently near to 0, there exists at least one $x \in V$ with $F_0(x) = c$. If there existed another distinct x' with $F_0(x') = c$, then we would have $x' \notin V$. Considering a sequence of c's that converge to 0, but for which distinct x and x' exist with $F_0(x) = F_0(x') = c$, we would conclude that the sequence of x's contained in V converges to x_0, while the other sequence of x's, or at least a subsequence of them, would converge to another distinct zero of F_0. This is a contradiction.

A similar argument shows that, for c in a sufficiently small neighborhood of $0 \in \mathbf{R}^N$, $H(t, x) \neq c$ holds for all $(t, x) \in [0, 1] \times \partial U$.

Now, by Sard's Theorem 4.2.7, the set of critical values $c \in \mathbf{R}^N$ is of Lebesgue N-dimensional measure zero. Therefore, for almost every choice of c near enough to 0, there is a unique x_c with $F_0(x_c) = c$, the matrix

$$\left(\frac{\partial H_i}{\partial x_j}(0, x_0) \right)_{i,j=1,2,\ldots,N}$$

is nonsingular, and there is no point in $[0, 1] \times \partial U$ at which H equals c. The result now follows by applying Corollary 4.2.5 to $H(t, x) - c$. □

We have touched only briefly on the topic of numerical homotopy methods; there is a substantial literature on the subject and its applications. We direct the reader to Allgower and Georg [AG 90] for a fuller exposition and additional references.

4.3 Equivalent Definitions of a Smooth Surface

In studying geometric analysis, one is often interested in considering a smooth surface in Euclidean space. We may all believe that we understand intuitively what the term "smooth surface in Euclidean space" means, but, of course, a precise definition must be given. In fact, there are many acceptable definitions for "smooth surface in Euclidean space." In this section, we will consider five equivalent definitions that correspond to the following five heuristic descriptions of a smooth surface:

- The surface can be smoothly straightened.
- The surface solves a system of smooth equations.
- The surface can be smoothly parametrized.
- The surface has smooth local coordinates.
- The surface is the graph of a smooth function.

The following theorem gives a precise expression to each of these heuristic descriptions. Of course, the point of the theorem is that the five possible definitions are equivalent, so any one of the five can be used to define "smooth surface in Euclidean space." To distinguish the precise from the intuitive, we then introduce the technical term *regularly imbedded C^k submanifold of* $\mathbf{R}^N$ in Definition 4.3.2.

Theorem 4.3.1 *Let M, N be integers with $1 \le M < N$. Let k be equal either to an integer greater than or equal to 1 or to $+\infty$. Let S be a subset of $\mathbf{R}^N$. The following are equivalent:*

(1) [The surface can be smoothly straightened.] *For each point $p \in S$ there exist a neighborhood $U \subset \mathbf{R}^N$ of p, a C^k diffeomorphism $\phi : U \to \mathbf{R}^N$, and an M dimensional linear subspace $L \subset \mathbf{R}^N$ such that*

$$\phi(S \cap U) = L \cap \phi(U). \tag{4.15}$$

(2) [The surface solves a system of smooth equations.] *For each point $p \in S$ there exist a neighborhood $U \subset \mathbf{R}^N$ of p and a C^k function $f : U \to \mathbf{R}^{N-M}$ such that,*

$$S \cap U = f^{-1}(q), \tag{4.16}$$

where $q = f(p)$, and

$$\operatorname{rank} Df(x) = N - M \tag{4.17}$$

holds for all $x \in U$.

(3) [The surface can be smoothly parametrized.] *For each $p \in S$ there exist an open $V \subset \mathbf{R}^M$ and a C^k function $g : V \to \mathbf{R}^N$ such that g maps each open subset of V onto a relatively open subset of S, p is in the image of g, and*

$$\operatorname{rank} Dg(x) = M \tag{4.18}$$

holds for all $x \in V$.

(4) [The surface has smooth local coordinates.] *For each point $p \in S$ there exist a neighborhood $U \subset \mathbf{R}^N$ of p, a convex open $W \subset \mathbf{R}^M$, and C^k functions $\phi : U \to W$, $\psi : W \to U$ such that*

$$S \cap U = \psi(W) \tag{4.19}$$

and $\phi \circ \psi$ is the identity map.

(5) [The surface is the graph of a smooth function.] *For each point $p \in S$ there exist a neighborhood $V \subset \mathbf{R}^N$ of p such that $S \cap V$ is the graph of a C^k function. More precisely, there exists a permutation of coordinates $\Phi : \mathbf{R}^N \to \mathbf{R}^N$, an open set $\mathcal{U} \subseteq \mathbf{R}^M$, and a C^k function $F : \mathcal{U} \to \mathbf{R}^{N-M}$, with $DF(u)$ of rank M for all $u \in \mathcal{U}$, such that*

$$S \cap V = \Phi\Big(\big\{(u, F(u)) : u \in \mathcal{U} \cap \mathbf{R}^M\big\}\Big). \tag{4.20}$$

Definition 4.3.2 If S satisfies any or, equivalently, all of the conditions of Theorem 4.3.1, then we say that S is an *M-dimensional, regularly imbedded C^k submanifold of $\mathbf{R}^N$*.

Proof of Theorem 4.3.1. The implicit function theorem and the inverse function theorem are the main tools used in showing the equivalence of the statements.

(1) $\Rightarrow$ (2) Let U, ϕ, and L be as in (1). Choose an orthonormal set of $N - M$ vectors $v_1, v_2, \ldots, v_{N-M}$, all orthogonal to L. We define $f : U \to \mathbf{R}^{N-M}$ by setting

$$f(x) = \Big(v_1 \cdot \phi(x), v_2 \cdot \phi(x), \ldots, v_{N-M} \cdot \phi(x)\Big).$$

Clearly, the function f is C^k. We see from (4.15) that $f(p) = 0$ and $S \cap U = f^{-1}(0)$. Because ϕ is a diffeomorphism, the rank of Df must be $N - M$ everywhere in U.

(2) ⇒ (3) Let U, f, and q be as in (2). Since rank $Df(x) = N - M$ holds for all $x \in U$, we can select indices $1 \le k_1 < k_2 < \cdots < k_{N-M} \le N$ so that

$$\det \begin{pmatrix} \frac{\partial f_1}{\partial x_{k_1}} & \frac{\partial f_1}{\partial x_{k_2}} & \cdots & \frac{\partial f_1}{\partial x_{k_{N-M}}} \\ \frac{\partial f_2}{\partial x_{k_1}} & \frac{\partial f_2}{\partial x_{k_2}} & \cdots & \frac{\partial f_2}{\partial x_{k_{N-M}}} \\ \vdots & \vdots & & \vdots \\ \frac{\partial f_{N-M}}{\partial x_{k_1}} & \frac{\partial f_{N-M}}{\partial x_{k_2}} & \cdots & \frac{\partial f_{N-M}}{\partial x_{k_{N-M}}} \end{pmatrix} \neq 0 \tag{4.21}$$

holds at p.

Let $1 \le j_1 < j_2 < \cdots < j_M \le N$ be the complementary indices so that

$$\{j_1, j_2, \ldots, j_M\} \cup \{k_1, k_2, \ldots, k_{N-M}\} = \{1, 2, \ldots, N\}.$$

Set

$$\xi_1 = x_{j_1},\ \xi_2 = x_{j_2}, \ldots, \xi_M = x_{j_M},$$

$$\eta_1 = x_{k_1},\ \eta_2 = x_{k_2}, \ldots, \eta_{N-M} = x_{k_{N-M}},$$

and

$$\widetilde{f}(\xi_1, \xi_2, \ldots, \xi_M, \eta_1, \eta_2, \ldots, \eta_{N-M}) = f(x_1, x_2, \ldots, x_N) - q.$$

Then we can apply the implicit function theorem to the function $\widetilde{f}$ at the point $(p_{j_1}, p_{j_2}, \ldots, p_{j_M}, p_{k_1}, p_{k_2}, \ldots, p_{k_{N-M}})$ to conclude that there exist a neighborhood $V \subset \mathbf{R}^M$ of $(p_{j_1}, p_{j_2}, \ldots, p_{j_M})$ and a C^k function $\widetilde{g} : V \to \mathbf{R}^{N-M}$ so that

$$\widetilde{f}(\xi, \widetilde{g}(\xi)) = 0$$

holds for all $\xi \in V$. Defining $g : V \to \mathbf{R}^N$ by setting

$$g_{j_1}(\xi) = \xi_1,\ g_{j_2}(\xi) = \xi_2, \ldots, g_{j_M}(\xi) = \xi_M$$

and

$$g_{k_1}(\xi) = \widetilde{g}_1(\xi),\ g_{k_2}(\xi) = \widetilde{g}_2(\xi), \ldots, g_{k_{N-m}}(\xi) = \widetilde{g}_{N-M}(\xi),$$

we see that the conditions in (3) hold.

(3) ⇒ (4) Let $V \subset \mathbf{R}^M$ and $g : V \to \mathbf{R}^N$ be as in (3). Let $v \in V$ be such that $g(v) = p$. Since rank $Dg(x) = M$ holds for all $x \in V$, we can choose indices $1 \le i_1 < i_2 < \cdots < i_M \le N$ so that

$$\det \begin{pmatrix} \frac{\partial g_{i_1}}{\partial x_1} & \frac{\partial g_{i_1}}{\partial x_2} & \cdots & \frac{\partial g_{i_1}}{\partial x_M} \\ \frac{\partial g_{i_2}}{\partial x_1} & \frac{\partial g_{i_2}}{\partial x_2} & \cdots & \frac{\partial g_{i_2}}{\partial x_M} \\ \vdots & \vdots & & \vdots \\ \frac{\partial g_{i_M}}{\partial x_1} & \frac{\partial g_{i_M}}{\partial x_2} & \cdots & \frac{\partial g_{i_M}}{\partial x_M} \end{pmatrix} \neq 0 \tag{4.22}$$

holds at v.

By the inverse function theorem, there exists an open set W with $v \in W \subseteq V$ on which the map $\hat{g}$ defined by setting

$$\hat{g}(x) = (g_{i_1}(x), g_{i_2}(x), \ldots, g_{i_M}(x))$$

is one-to-one and has a C^k inverse. Without loss of generality, we may assume that W is convex.

Set

$$U = \left\{(u_1, u_2, \ldots, u_N) : (u_{i_1}, u_{i_2}, \ldots, u_{i_M}) \in \hat{g}(W)\right\},$$

and define $\phi : U \to W$ by setting

$$\phi(u_1, u_2, \ldots, u_N) = \hat{g}^{-1}(u_{i_1}, u_{i_2}, \ldots, u_{i_M}).$$

We see that the conditions of (4) hold with $\psi = g|_W$.

(4) $\Rightarrow$ (5) Let U, W, ϕ, and ψ be as in (4). Since $\phi \circ \psi$ is the identity, $D\psi$ must be of rank M at each point of W, in particular, at $q = \phi(p)$. Thus we can choose indices $1 \leq i_1 < i_2 < \cdots < i_M \leq N$ so that

$$\det \begin{pmatrix} \frac{\partial \psi_{i_1}}{\partial w_1} & \frac{\partial \psi_{i_1}}{\partial w_2} & \cdots & \frac{\partial \psi_{i_1}}{\partial w_M} \\ \frac{\partial \psi_{i_2}}{\partial w_1} & \frac{\partial \psi_{i_2}}{\partial w_2} & \cdots & \frac{\partial \psi_{i_2}}{\partial w_M} \\ \vdots & \vdots & & \vdots \\ \frac{\partial \psi_{i_M}}{\partial w_1} & \frac{\partial \psi_{i_M}}{\partial w_2} & \cdots & \frac{\partial \psi_{i_M}}{\partial w_M} \end{pmatrix} \neq 0 \tag{4.23}$$

holds at q.

Let $1 \leq j_1 < j_2 < \cdots < j_{N-M} \leq N$ be the complementary indices so that

$$\{i_1, i_2, \ldots, i_M\} \cup \{j_1, j_2, \ldots, j_{N-M}\} = \{1, 2, \ldots, N\}.$$

Define the orthogonal transformation $\Psi : \mathbf{R}^N \to \mathbf{R}^N$ by setting

$$\Psi(x_1, x_2, \ldots, x_N) = (x_{i_1}, x_{i_2}, \ldots, x_{i_M}, x_{j_1}, x_{j_2}, \ldots, x_{j_{N-M}}).$$

Let $\Pi_1 : \mathbf{R}^M \times \mathbf{R}^{N-M} \to \mathbf{R}^M$ and $\Pi_2 : \mathbf{R}^M \times \mathbf{R}^{N-M} \to \mathbf{R}^{N-M}$ be the projections onto the first and second factors, respectively. By (4.23), we can apply the inverse function theorem to $\Pi_1 \circ \Psi \circ \psi$ at the point $u_0 = \Pi_1 \circ \Psi \circ \psi(q)$ to see that

$$f = \left(\Pi_1 \circ \Psi \circ \psi\right)^{-1} : \mathcal{U} \subseteq \mathbf{R}^M \to \mathbf{R}^M$$

exists, is C^k, and Df has rank M at each point $u \in \mathcal{U}$. Setting

$$F = \Pi_2 \circ \Psi \circ \psi \circ f, \qquad \Phi = \Psi^{-1},$$

and taking $\mathcal{U}$ to be a possibly smaller neighborhood of u_0, we obtain (4.20).

(5) $\Rightarrow$ (1) Let V, Φ, and $\mathcal{U}$ be as in (5). Let $\Pi_1 : \mathbf{R}^M \times \mathbf{R}^{N-M} \to \mathbf{R}^M$ and $\Pi_2 : \mathbf{R}^M \times \mathbf{R}^{N-M} \to \mathbf{R}^{N-M}$ be the projections onto the first and second factors, respectively. Let $u_0 \in \mathcal{U}$ be such that $p = \Phi(u_0, F(u_0))$, and let L be the image under Φ of the tangent plane to the graph of F at $(u_0, F(u_0))$.

We define $H = \Pi_1 \circ \Phi^{-1}$ and $V = \Pi_2 \circ \Phi^{-1} - F \circ H$. Then, defining $\phi : \mathbf{R}^N \to \mathbf{R}^N$ by setting[2]

$$\phi(x) = \Phi\Big(H(x), \; V(x) + F(u_0) + \langle DF(u_0), \Pi_1 \circ \Phi^{-1}(x) - u_0\rangle\Big),$$

we see that (4.15) holds for a small enough neighborhood of p. □

We close with a remark about the significance of the phrase "regularly imbedded." In the geometric theory of manifolds, one learns that a manifold is a geometric object that is locally equivalent to Euclidean space—the equivalence provided by a smooth mapping. The definition of manifold is *independent of any imbedding*. And it is entirely possible to map a smooth manifold into Euclidean space in a nonsmooth way. For example, the unit interval (which is certainly a smooth manifold) can be mapped into the plane in a continuous and one-to-one manner so that the image has positive area (see Osgood [Os 03]). Such a curve is not even rectifiable, hence is certainly not smooth.

When a smooth mapping of a manifold into Euclidean space is called an "imbedding," it is always required that the mapping be one-to-one, so self-intersections are not allowed. Even in the absence of self-intersections, it *may or may not be true* that the following property holds:

> The mapping is a homeomorphism onto its image, when the image has the relative topology.

Some authors require this stronger homeomorphism property of every mapping that they call an imbedding. In our lexicon, we use the term "regular imbedding" when demanding the homeomorphism property. As we see from Theorem 4.3.1, a regularly imbedded manifold sits in space in a smooth and geometrically natural manner, much as the line $\{(x, y) : y = 0\}$ sits in the plane $\mathbf{R}^2$.

---

[2]Recall from Section 3.3 that we use the notation $\langle\ ,\ \rangle$ to denote the application of the Jacobian matrix to a vector.

4.4 Smoothness of the Distance Function

In the study of various problems of analysis that may be set in a bounded domain in Euclidean space, it is often useful, or even necessary, to consider the distance from a point to the boundary of the domain. Of course one can also consider the distance to the boundary of a domain when measured from a point *outside* the domain. Assigning a positive sign to the distance to the boundary from the inside and a negative sign to the distance to the boundary from the outside, one is led to the idea of the signed distance function to a surface.

In what follows we will not always refer to the inside and outside of a domain because we can equivalently use a choice of unit normal vector to orient a surface.

It may be noted that, in general, the ordinary distance will not be a smooth function. For example, let S be a line in $\mathbf{R}^2$. Then the function

$$\varrho(x) \equiv \inf\{|s - x| : x \in S\}$$

is not even C^1. One of the main motivations for considering the signed distance function is that it is maximally smooth under minimal hypotheses.

Let us assume for purposes of a motivating discussion that the surface is smooth and that we are investigating the signed distance function in a region near enough to the surface that, for each point in the region, there is a unique nearest point of the surface.[3] The most straightforward way to measure the distance from a point p to the surface is to locate the nearest point $\xi(p)$ on the surface and compute the length of the vector from the nearest point to the original point. To get the signed distance function, one uses the fact that the vector from $\xi(p)$ to p must be perpendicular to the surface at $\xi(p)$. To take advantage of this geometry, one chooses a continuous unit normal field on the surface and computes the inner product of that unit normal at that nearest point $\xi(p)$ with the vector from the nearest point to the original point.

In this process of finding the nearest point on the surface and taking the inner product with the unit normal vector, it seems that at least one differentiation has taken place and that the signed distance function would be expected to be one order of differentiability less smooth than the surface. In fact, the signed distance function is just as smooth as the original surface. The key to proving this is in making efficient use of the implicit function theorem or of the inverse function theorem. As far as we know, this result first appeared[4] in Gilbarg and Trudinger [GT 77; Lemma 1, page 382].[5] Other proofs can be found in Krantz and Parks [KP 81], Foote [Fo 84], and Krantz and Parks [KP 99; Theorem 1.2.6,

---

[3]In fact, this is a nontrivial hypothesis. A surface with this property is called a "set of positive reach." It is known, and we will prove below using the implicit function theorem, that a C^2 surface is a set of positive reach.

[4]On page 50 of Hörmander [Ho 66] it is observed in passing that the implicit function theorem can be used to show the distance function is as smooth as the boundary.

[5]In the second edition, Gilbarg and Trudinger [GT 83], the result appears as Lemma 14.16, page 355.

page 12]. Lars Hörmander was good enough to describe to us a proof using the calculus of variations [Ho 00].

General Facts about the Distance Function

In the proof of the next proposition we show that, for an arbitrary closed set, it follows from the triangle inequality that the distance function is a Lipschitz function with Lipschitz constant 1.

Proposition 4.4.1 *Let S be a closed subset of $\mathbf{R}^N$. Then for any $x, y \in \mathbf{R}^N$ it holds that*

$$|\operatorname{dist}(x, S) - \operatorname{dist}(y, S)| \le |x - y| .$$

Proof. Let $x, y \in \mathbf{R}^N$ be arbitrary points and let $v \in S$ be such that

$$\operatorname{dist}(y, S) = |y - v|$$

holds. Then we have

$$\operatorname{dist}(x, S) \le |x - v| \le |x - y| + |y - v| ,$$

so

$$\operatorname{dist}(x, S) - \operatorname{dist}(y, S) \le |x - y|$$

holds. Similarly, we have

$$\operatorname{dist}(y, S) - \operatorname{dist}(x, S) \le |x - y| ,$$

proving the result. □

To learn more about the distance function we will need to investigate the behavior (on its domain) of the unique nearest point function. First, we give the definition.

Definition 4.4.2 Let S be a closed subset of $\mathbf{R}^N$ and denote by $\Xi(S)$ the set of $x \in \mathbf{R}^N$ such that

$$\xi_1 \in S,\ \xi_2 \in S, \text{ and } |x - \xi_1| = |x - \xi_2| = \operatorname{dist}(x, S) \text{ imply } \xi_1 = \xi_2 .$$

Define the *nearest point function* $\xi : \Xi(S) \to S$ by requiring

$$\xi(x) \in S \text{ and } |x - \xi(x)| = \operatorname{dist}(x, S),$$

for $x \in \Xi(S)$. Of course, $\Xi(S)$ is exactly the set on which the nearest point function ξ is well defined.

In the proof of the next lemma we will see that it is elementary to show that, for an arbitrary closed set, the unique nearest point is continuous where it is defined.

Lemma 4.4.3 *For any closed $S \subset \mathbf{R}^N$, the function $\xi : \Xi(S) \to S$ is continuous.*

Proof. Arguing by contradiction, suppose that $x_1, x_2, \ldots$ is a sequence in $\Xi(S)$ that converges to $x \in \Xi(S)$, but that there is $\epsilon > 0$ such that $|\xi(x_i) - \xi(x)| \geq \epsilon$ holds for all $i = 1, 2, \ldots$.

It is clear that $\xi(x_1), \xi(x_2), \ldots$ is a bounded sequence so, by passing to a subsequence if necessary, but without changing notation, we may assume that $\xi(x_i)$ converges to some point $z \in S$. It is also clear that $|z - x| = \operatorname{dist}(x, S)$, so by the definition of $\Xi(S)$ we must have $\xi(x) = z$. contradicting the assumption that $\xi(x_i)$ stays at least a distance of ϵ away from $\xi(x)$. □

Even for a completely arbitrary subset of Euclidean space, the directional derivatives of the distance function can often be shown to behave well, where they exist. The next two lemmas appear in Federer [Fe 59].

Lemma 4.4.4 *For any closed $S \subset \mathbf{R}^N$, let the function $\varrho : \mathbf{R}^N \to \mathbf{R}$ be defined by setting*

$$\varrho(x) = \operatorname{dist}(x, S).$$

If $x \in \Xi(S) \setminus S$ and if the directional derivative of ϱ in the direction of the unit vector v exists, then that directional derivative equals

$$v \cdot \frac{x - \xi(x)}{\varrho(x)}. \tag{4.24}$$

Proof. Fix $x \in \Xi(S) \setminus S$ and write simply ξ for $\xi(x)$. Also write $r = |x - \xi|$.

Fix the unit vector v and suppose that the directional derivative

$$\lim_{t \to 0} \frac{\varrho(x + tv) - \varrho(x)}{t}$$

exists.

Since $\xi \in S$, we have

$$\begin{aligned} \varrho(x + tv) \leq |x + tv - \xi| &= \sqrt{(x + tv - \xi) \cdot (x + tv - \xi)} \\ &= \sqrt{(x - \xi) \cdot (x - \xi) + 2tv \cdot (x - \xi) + t^2 v \cdot v} \\ &= \sqrt{r^2 + 2tv \cdot (x - \xi) + t^2}. \end{aligned}$$

So we have

$$\begin{aligned} \lim_{t \downarrow 0} \frac{\varrho(x + tv) - \varrho(x)}{t} &\leq \lim_{t \downarrow 0} \frac{\sqrt{r^2 + 2tv \cdot (x - \xi) + t^2} - r}{t} \\ &= \lim_{t \downarrow 0} \frac{2v \cdot (x - \xi) + t}{\sqrt{r^2 + 2tv \cdot (x - \xi) + t^2} + r} \\ &= \frac{v \cdot (x - \xi)}{r}. \end{aligned}$$

Likewise, we have

$$\lim_{t\uparrow 0}\frac{\varrho(x+tv)-\varrho(x)}{t} \geq \lim_{t\uparrow 0}\frac{\sqrt{r^2+2tv\cdot(x-\xi)+t^2}-r}{t} = \frac{v\cdot(x-\xi)}{r},$$

and the result follows. □

Remark 4.4.5 By Lemma 4.4.3, the expression in (4.24) is continuous on $\Xi(S)\setminus S$. This observation will be used in the proof of the next theorem to conclude that the distance function is continuously differentiable on the interior of $\Xi(S)\setminus S$. If the closed set S under consideration is actually a C^1 submanifold, then one can make a direct argument based on examining the difference quotient to conclude that (4.24) gives the directional derivative at all points in the interior of $\Xi(S)\setminus S$. In this way the less elementary argument used in the proof of the next theorem can be avoided for a C^1 submanifold (see Foote [Fo 84; Theorem 2]).

Theorem 4.4.6 *For any closed $S\subset\mathbf{R}^N$, the function $\varrho:\mathbf{R}^N\to\mathbf{R}$ defined by setting*

$$\varrho(x)=\operatorname{dist}(x,S)$$

is continuously differentiable on the interior of $\Xi(S)\setminus S$.

Proof. Consider a point in the interior of $\Xi(S)\setminus S$ and the line L through that point and parallel to the ith coordinate axis. Since by Proposition 4.4.1 we have

$$|\varrho(x_1)-\varrho(x_2)|\leq|x_1-x_2|$$

for any pair of points $x_1,x_2\in\mathbf{R}^N$, it follows that ϱ is absolutely continuous on L and thus that ϱ is the integral of its derivative along L. But the derivative of ϱ along L is the ith partial derivative of ϱ and, by Lemma 4.4.4, where it exists it equals the continuous function

$$\mathbf{e}_i\cdot\frac{x-\xi(x)}{\varrho(x)}.$$

Here $\mathbf{e}_i$ denotes the ith standard basis vector for $\mathbf{R}^N$. Consequently, ϱ restricted to L intersected with the interior of $[\Xi(S)\setminus S]$ is the integral of a continuous function, and thus is continuously differentiable on L.

It follows that all the partial derivatives of ϱ exist and are continuous on the interior of $[\Xi(S)\setminus S]$, so ϱ is continuously differentiable on that set. □

The Distance to a Submanifold

Next we present a useful construction which, for a smooth submanifold of $\mathbf{R}^N$, gives a smooth parametrization of a tubular neighborhood about M near any point. The ideas here go back to Hotelling [Ho 39] and the seminal paper of Weyl [We 39]. For a thorough treatment of tubes, the reader should see Gray [Gr 90].

Lemma 4.4.7 (Local Tubular Neighborhood Lemma) *Let $M \subset \mathbf{R}^N$ be a compact C^k submanifold of dimension K, where we assume $k \geq 2$ and $N - 1 \geq K \geq 1$. Then, for each point P in M, there exist a neighborhood W of P in M, a C^k function $F : W' \to \mathbf{R}^N$, where $W' \subset \mathbf{R}^K$ is open, and an orthonormal set of C^{k-1} vector fields $V_j : W' \to \mathbf{R}^N$, $j = 1, 2, \ldots, N - K$, such that*

(1) $F(W') = W$,

(2) rank $(DF) = K$ *at every point of* W',

(3) $V_j(x)$ *is normal to* M *at* $F(x)$ *for each* $x \in W'$ *and for each* $j = 1, 2, \ldots, N - K$,

(4) $\Phi : W' \times \mathbf{R}^{N-K} \to \mathbf{R}^N$ *defined by*

$$\Phi(x, y) = F(x) + \sum_{j=1}^{N-K} y_j V_j(x), \tag{4.25}$$

for $x \in W'$ and $y = (y_1, y_2, \ldots, y_{N-K}) \in \mathbf{R}^{N-K}$, is a C^{k-1} function, is one-to-one on $\{x\} \times \mathbf{R}^{N-K}$ for each $x \in W'$, and is one-to-one in a neighborhood of $W' \times \{0\}$.

Proof. Because $M \subseteq \mathbf{R}^N$ is a C^k submanifold, we can choose a neighborhood W of P in M that is parametrized by

$$F : W' \subset \mathbf{R}^K \to \mathbf{R}^N$$

where rank $(DF) = K$ at every point of W'. Thus (1) and (2) are satisfied.

By taking a smaller neighborhood than W' if necessary, but without changing notation, we can arrange that *the same K rows*,

$$i_1, i_2, \ldots, i_K, \tag{4.26}$$

of the $N \times K$ matrix

$$\left(\frac{\partial F_i}{\partial x_j}\right)_{\substack{i=1,\ldots,N \\ j=1,\ldots,K}} \tag{4.27}$$

are independent at every point of W'. Now, because the same rows of the matrix (4.27) are independent at every point of W', we will be able to define a set of C^{k-1} vector fields

$$T_1(x), T_2(x), \ldots, T_K(x), \; V_1(x), V_2(x), \ldots, V_{N-K}(x),$$

for $x \in W'$, that form an orthonormal basis for $\mathbf{R}^N$ for each choice of $x \in W'$. We may further suppose that

$$T_1(x), T_2(x), \ldots, T_K(x),$$

are all tangent to M at $F(x)$, while

$$V_1(x), V_2(x), \ldots, V_{N-K}(x)$$

are all normal to M at $F(x)$. One way to do this is to apply the Gram–Schmidt orthogonalization procedure to the set of vectors consisting of

$$\left(\frac{\partial F}{\partial x_j}\right), \; j = 1, \ldots, K,$$

and the standard basis vectors

$$\mathbf{e}_{i'_1}, \mathbf{e}_{i'_2}, \ldots, \mathbf{e}_{i'_{N-K}},$$

where $i'_1, i'_2, \ldots, i'_{N-K}$ are the indices *not* occurring in (4.26). Thus, (3) is satisfied.

Because of (1) and (3), it is clear that Φ defined by (4.25) is a C^{k-1} function. Observe that $D\Phi(x, 0)$ is represented by the matrix that has as its first K columns the independent columns of $DF(x)$ that are tangent to M at $F(x)$, and that has as its last $N - K$ columns the vectors $V_1(x), V_2(x), \ldots, V_{N-K}(x)$ that are independent vectors normal to M at $F(x)$; thus we see that $D\Phi(x, 0)$ is nonsingular for each $x \in W'$. Finally, we apply the inverse function theorem to conclude that Φ is one-to-one in a neighborhood of $W' \times \{0\}$, as claimed in (4). □

The following definition was introduced in Federer [Fe 59]. It will allow us to generalize the preceding result.

Definition 4.4.8 A set $S \subseteq \mathbf{R}^N$ is said to be of *positive reach* if there exists $r > 0$ such that,

$$\text{for all } x \in \mathbf{R}^N, \; \text{dist}(x, S) < r \text{ implies there is a unique point } \xi \in S \text{ with } |x - \xi| = \text{dist}(x, S). \tag{4.28}$$

In case S is of positive reach, we define the *reach of* S to be the supremum of all $r > 0$ for which the condition (4.28) is true.

If one examines the proof of Lemma 4.4.7, one will note that the only place in which we used the fact that M was at least C^2 was to apply the inverse function theorem to conclude that Φ is one-to-one in a neighborhood of $W' \times \{0\}$. By adding the hypothesis that M is of positive reach, we can obtain the same conclusion in the C^1 case. These remarks give us the following corollary of the proof of Lemma 4.4.7.

Corollary 4.4.9 *If $M \subset \mathbf{R}^N$ is a compact C^1 submanifold of dimension K, where we assume $N-1 \geq K \geq 1$, and if M has positive reach, then for each point P in M there exist a neighborhood W of P in M, a C^k function $F : W' \to \mathbf{R}^N$, where $W' \subset \mathbf{R}^K$ is open, and an orthonormal set of C^{k-1} vector fields $V_j : W' \to \mathbf{R}^N$, $j = 1, 2, \ldots, N-K$, such that*

(1) $F(W') = W$,

(2) $\operatorname{rank}(DF) = K$ *at every point of* W'

(3) $V_j(x)$ *is normal to* M *at* $F(x)$ *for each* $x \in W'$ *and for each* $j = 1, 2, \ldots, N-K$,

(4) $\Phi : W' \times \mathbf{R}^{N-K} \to \mathbf{R}^N$ *defined by*

$$\Phi(x, y) = F(x) + \sum_{j=1}^{N-K} y_j\, V_j(x), \tag{4.29}$$

for $x \in W'$ *and* $y = (y_1, y_2, \ldots, y_{N-K}) \in \mathbf{R}^{N-K}$, *is a* C^{k-1} *function, is one-to-one on* $\{x\} \times \mathbf{R}^{N-K}$ *for each* $x \in W'$, *and is one-to-one in a neighborhood of* $W' \times \{0\}$.

Theorem 4.4.10 *If $M \subset \mathbf{R}^N$ is a compact C^k submanifold of dimension K, where $k \geq 2$ and $N-1 \geq K \geq 1$, then M has positive reach and there is a neighborhood U of M on which $\xi : U \to M$ is C^{k-1}.*

Proof. Suppose that M is not of positive reach. Then there exist sequences of points $X_1, X_2, \ldots$ in $\mathbf{R}^N$, $P_1, P_2, \ldots$ in M, and $Q_1, Q_2, \ldots$ in M with

$$0 = \lim_{i \to \infty} \operatorname{dist}(X_i, M), \tag{4.30}$$

and such that, for each $i = 1, 2, \ldots,$

$$P_i \neq Q_i \text{ and } |X_i - P_i| = |X_i - Q_i| = \operatorname{dist}(X_i, M) \tag{4.31}$$

hold. By the compactness of M, we can pass to a subsequence, but without changing notation, so that

$$\lim_{i \to \infty} X_i = \lim_{i \to \infty} P_i = \lim_{i \to \infty} Q_i = P \in M. \tag{4.32}$$

We apply Lemma 4.4.7 at P to obtain the neighborhood W of P in M, the function $F : W' \to \mathbf{R}^N$, and the function $\Phi : W' \times \mathbf{R}^{N-K} \to \mathbf{R}^N$. Letting $x_0 \in W'$ be such that $F(x_0) = P$, we can use Lemma 4.4.6(4) to choose a neighborhood U_P of P in $\mathbf{R}^N$ so that Φ is one-to-one on U_P.

By (4.32), for all large enough i, we have $P_i, Q_i, X_i \in U_P$ contradicting the fact that Φ is one-to-one there. Thus M is of positive reach.

Note that on U_P the nearest point retraction is $F \circ \Pi \circ \Phi^{-1}$, where Π is projection onto the first factor. Therefore ξ is a C^{k-1} function on U_P. Since the same reasoning can be applied at *any* point $P \in M$, we see that ξ is C^{k-1} in a neighborhood of M. □

Corollary 4.4.11 (Foote [Fo 84]) *If $M \subset \mathbf{R}^N$ is a compact C^k submanifold of dimension K, where $k \geq 2$ and $N - 1 \geq K \geq 1$, then there is a neighborhood U of M on which the distance function*

$$\varrho(x) = \text{dist}(x, M)$$

is C^k on $U \setminus M$.

Proof. By Theorem 4.4.6, the distance function is continuously differentiable on the interior of $\Xi(M) \setminus M$. Any directional derivative of the distance function is given by Equation (4.24) in Lemma 4.4.3, but by Theorem 4.4.10, that directional derivative is a C^{k-1} function on $U \setminus M$. Thus, ϱ itself is C^k. □

The following example shows us that, for a surface that is less smooth than C^2, everything can go wrong: The surface need not be of positive reach and the distance function can fail to be differentiable at points arbitrarily near to the surface.

Example 4.4.12 Let $0 < \epsilon < 1$ be fixed. Form a simple closed curve γ in $\mathbf{R}^2$ by smoothly connecting the endpoints of the graph of

$$y = |x|^{2-\epsilon}, \quad -1 \leq x \leq 1. \tag{4.33}$$

Consider $0 < b$ small enough that any point (x_0, y_0) such that $\text{dist}[(0, b), \gamma] = |(0, b) - (x_0, y_0)|$ is a point on the graph (4.33). Notice that for small $|x| > 0$ it holds that

$$|x|^{\epsilon} + |x|^{2-\epsilon} < 2b,$$

so

$$|x|^{2-\epsilon}\left(|x|^{\epsilon} + |x|^{2-\epsilon}\right) + b^2 < 2|x|^{2-\epsilon}b + b^2.$$

Thus, for any such small $|x| > 0$, we have

$$|x|^2 + \left(|x|^{2-\epsilon}\right)^2 - 2|x|^{2-\epsilon}b + b^2 < b^2$$

or

$$\left|(0, b) - \left(x, |x|^{2-\epsilon}\right)\right| < |(0, b) - (0, 0)|. \tag{4.34}$$

Because of (4.34), we conclude that $\text{dist}[(0, b), \gamma] < |(0, b) - (0, 0)|$ and by the symmetry of the graph (4.33) there must be at least two points $\left(x_0, |x_0|^{2-\epsilon}\right)$ and $\left(-x_0, |x_0|^{2-\epsilon}\right)$, where $x_0 > 0$, for which

$$\text{dist}[(0, b), \gamma] = \left|(0, b) - \left(x_0, |x_0|^{2-\epsilon}\right)\right| = \left|(0, b) - \left(-x_0, |x_0|^{2-\epsilon}\right)\right|$$

holds.

Now, at any point such as $(0, b)$ above for which there are two or more distinct nearest points, the distance function must fail to be differentiable. This is so because a function that is differentiable at a point and for which the gradient does not vanish has a well-defined direction of most rapid decrease, but when there are two distinct nearest points, there are two corresponding directions of most rapid decrease for the distance function.

In fact, the situation is even worse in this example. By taking $\epsilon = 1/2$ for simplicity and taking b small enough as above, one can compute (letting δ denote distance to the curve) that

$$\lim_{h\downarrow 0}\frac{\delta[(h,b)]-\delta[(0,b)]}{h}\neq\lim_{h\uparrow 0}\frac{\delta[(h,b)]-\delta[(0,b)]}{h},$$

so $\partial\delta/\partial x$ does not exist at $(0, b)$. □

The Generalized Distance Function

In our results concerning the smoothness of the distance function, we have so far rather conspicuously avoided the points of M itself. In fact, the (unsigned) distance function, without some modification, must fail to be differentiable at all points of M. The surprising thing is that in the cases $K = N - 1$ and $K = 1$, it is possible to modify the distance function ϱ so as to make the resulting function $\tilde{\varrho}$ still be as smooth as the manifold M, but on an entire neighborhood of M. To achieve this end, we need an improved form of the Local Tubular Neighborhood Lemma 4.4.7. As far as we know, the key part of this improved result, (2), can only be obtained for curves, $K = 1$, and hypersurfaces, $K = N - 1$, and not for the intervening dimensions, $1 < K < N - 1$, of the submanifold.

Lemma 4.4.13 *If $M \subset \mathbf{R}^N$ is a compact C^k submanifold of dimension K equal to either 1 or $N - 1$, where $k \geq 2$ and $N \geq 2$, then for each point P in M there exist a neighborhood W of P in M, a C^k function $F : W' \to \mathbf{R}^N$, where $W' \subset \mathbf{R}^K$ is open, and an orthonormal set of C^{k-1} vector fields $\widetilde{V}_j : W' \to \mathbf{R}^N$, $j = 1, 2, \ldots, N - K$, such that*

(1) *$\widetilde{V}_j(x)$ is normal to M at $F(x)$ for each $x \in W'$ and for each $j = 1, 2, \ldots, N - K$,*

(2) *for each $P \in M$ and for each $j = 1, 2, \ldots, N - K$, the directional derivative of $\widetilde{V}_j$ in any direction tangent to M is again tangent to M,*

(3) *$\Phi : W' \times \mathbf{R}^{N-K} \to \mathbf{R}^N$ defined by*

$$\Phi(x,y) = x + \sum_{j=1}^{N-K} y_j\,\widetilde{V}_j(x), \tag{4.35}$$

for $x \in W'$ and $y = (y_1, y_2, \ldots, y_{N-K}) \in \mathbf{R}^{N-K}$, is a C^{k-1} function, is one-to-one on $\{x\} \times \mathbf{R}^{N-1}$ for each $x \in W'$, and is one-to-one in a neighborhood of $W' \times \{0\}$.

Proof. Proceed as in the proof of Lemma 4.4.7 to obtain W, F, and the orthonormal vector fields $V_j : W' \to \mathbf{R}^N$, $j = 1, 2, \ldots, N - K$. Suppose without loss of generality that $0 \in W'$ and $F(0) = P$.

In case $K = N - 1$, the condition (2) follows from the fact that $V_N \cdot V_N \equiv 1$; more precisely, differentiation of both sides of the equation $V_N \cdot V_N = 1$ in a tangent direction implies that the derivative of V_N in that direction must be orthogonal to V_N, and hence that directional derivative of V_N is tangent to M.

Now, assume that $K = 1$. We apply the Fundamental Existence Theorem for ordinary differential equations to find a set of functions

$$\begin{array}{ccc} \varphi_{1\,1}, \varphi_{1\,2}, & \ldots, & \varphi_{1\,N-1}, \\ \varphi_{2\,1}, \varphi_{2\,2}, & \ldots, & \varphi_{2\,N-1}, \\ & \ldots, & \\ \varphi_{N-1\,1}, \varphi_{N-1\,2}, & \ldots & \varphi_{N-1\,N-1}, \end{array}$$

satisfying

$$\begin{aligned} \varphi'_{i\,j} \;&+\; \left(1+\varphi_{j\,j}\right)^{-1} \sum_{\substack{\ell=1 \\ \ell \neq j}}^{N-1} \varphi_{j\,\ell}\varphi'_{i\,\ell} \\ &+\; \left(1+\varphi_{j\,j}\right)^{-1} \sum_{\ell,m=1}^{N-1} \left(\delta_{i\,\ell}+\varphi_{i\,\ell}\right)\left(\delta_{j\,m}+\varphi_{j\,m}\right) V'_\ell \cdot V_m = 0, \end{aligned} \tag{4.36}$$

$$\varphi_{i\,j}(0) = 0, \tag{4.37}$$

for $i = 1, 2, \ldots, N-1$ and $j = 1, 2, \ldots, N-1$. Here δ_{ij} is the standard Kronecker delta. Noting that the first-order differential equations in (4.36) involve data that are C^{k-2}, we conclude that the solutions $\varphi_{i\,j}$ are C^{k-1}. Replacing W and W' by smaller neighborhoods if necessary, but without changing notation, we may assume the functions $\varphi_{i\,j}$ are defined and satisfy (4.36) on all of W.

Set

$$\widetilde{V}_i(x) = \sum_{\ell=1}^{N-1} \left(\delta_{i\,\ell} + \varphi_{i\,\ell}(x)\right) V_\ell(x).$$

The vector fields $\widetilde{V}_i$ are C^{k-1}, because the $\varphi_{i\,\ell}$ and the V_ℓ are C^{k-1}. By (4.36), we have $\widetilde{V}_i(0) = V_i(0)$, so at $x = 0$ the system of vectors $\widetilde{V}_1(0), \widetilde{V}_2(0), \ldots, \widetilde{V}_{N-1}(0)$ is orthonormal. We compute

$$\begin{aligned} \widetilde{V}'_i \cdot \widetilde{V}_j &= \left(\sum_{\ell=1}^{N-1} \left(\delta_{i\,\ell}+\varphi_{i\,\ell}\right) V_\ell\right)' \cdot \left(\sum_{m=1}^{N-1} \left(\delta_{j\,m}+\varphi_{j\,m}\right) V_m\right) \\ &= \left(\sum_{\ell=1}^{N-1} \varphi'_{i\,\ell} V_\ell + \sum_{\ell=1}^{N-1} \left(\delta_{i\,\ell}+\varphi_{i\,\ell}\right) V'_\ell\right) \cdot \left(\sum_{m=1}^{N-1} \left(\delta_{j\,m}+\varphi_{j\,m}\right) V_m\right) \\ &= \sum_{\ell=1}^{N-1} \varphi'_{i\,\ell} V_\ell \cdot \left(V_j + \sum_{m=1}^{N-1} \varphi_{j\,m} V_m\right) \end{aligned}$$

$$+ \sum_{\ell,m=1}^{N-1} \left(\delta_{i\ell} + \varphi_{i\ell}\right)\left(\delta_{jm} + \varphi_{jm}\right) V'_\ell \cdot V_m$$

$$= \left(1 + \varphi_{jj}\right)\varphi'_{ij} + \sum_{\substack{\ell=1 \\ \ell \neq j}}^{N-1} \varphi_{j\ell}\varphi'_{i\ell}$$

$$+ \sum_{\ell,m=1}^{N-1} \left(\delta_{i\ell} + \varphi_{i\ell}\right)\left(\delta_{jm} + \varphi_{jm}\right) V'_\ell \cdot V_m$$

$$= 0$$

It follows that $\widetilde{V}_1(x), \widetilde{V}_2(x), \ldots, \widetilde{V}_{N-1}(x)$ is an orthonormal system for all $x \in W'$ and that (2) holds as required.

The remainder of the proof proceeds as for Lemma 4.4.7. □

Definition 4.4.14 Suppose $M \subset \mathbf{R}^N$ is a compact C^k submanifold of dimension either 1 or $N - 1$, where $k \geq 2$ and $N \geq 2$. Let the C^{k-1} function $\Phi : W' \times \mathbf{R}^{N-K} \to \mathbf{R}^N$ be as in Lemma 4.4.13(3). By the inverse function theorem, there is a neighborhood U of W on which Φ is invertible. Define $\tilde{\varrho} : U \to \mathbf{R}^{N-K}$ by setting

$$\tilde{\varrho}(Q) = \Pi_2 \circ \Phi^{-1}(Q), \tag{4.38}$$

for $Q \in U$, where $\Pi_2 : W' \times \mathbf{R}^{N-K} \to \mathbf{R}^{N-K}$ is projection onto the second factor.

Theorem 4.4.15 *If $M \subset \mathbf{R}^N$ is a compact C^k submanifold of dimension K equal to either 1 or $N - 1$, where $k \geq 2$ and $N \geq 2$, then $\tilde{\varrho}$ defined in (4.38) is C^k and satisfies*

$$|\tilde{\varrho}(Q)| = \varrho(Q), \tag{4.39}$$

for $Q \in U$, where U is as in Definition 4.4.14.

Remark 4.4.16

(1) In case $K = N - 1$, the function $\tilde{\varrho}$ defined in (4.38) is called the *signed distance function*. In case $K = 1$, we can consider $\tilde{\varrho}$ to be the generalization of the signed distance function, as is justified by (4.39).

(2) If the function Φ from Lemma 4.4.7 were used in Definition 4.4.14, instead of Φ from Lemma 4.4.13, then equation (4.39) would still hold, but our proof that $\tilde{\varrho}$ is C^k would no longer be valid.

Proof of Theorem 4.4.14. For each $j = 1, 2, \ldots, N - K$, define $\tilde{\varrho}_j : U \to \mathbf{R}$ by setting

$$\tilde{\varrho}_j(x) = \left[x - \xi(x)\right] \cdot \widetilde{V}_j[\xi(x)].$$

Using $\mathbf{I}$ to denote the identity map on $\mathbf{R}^N$, we compute[6]

$$\begin{aligned}
\left\langle D\widetilde{Q}_j(x), \widetilde{V}_\ell[\xi(x)]\right\rangle &= \left\langle \mathbf{I} - D\xi(x), \widetilde{V}_\ell[\xi(x)]\right\rangle \cdot \widetilde{V}_j[(\xi(x)] \\
&\quad + [x - \xi(x)] \cdot \left\langle D(\widetilde{V}_j \circ \xi)(x), \widetilde{V}_\ell[(\xi(x)]\right\rangle \\
&= \widetilde{V}_\ell[\xi(x)] \cdot \widetilde{V}_j[\xi(x)] && (4.40) \\
&\quad - \left\langle D\xi(x), \widetilde{V}_\ell[\xi(x)]\right\rangle \cdot \widetilde{V}_j[\xi(x)] && (4.41) \\
&\quad + [x - \xi(x)] \cdot \left\langle D\widetilde{V}_j[\xi(x)], \left\langle D\xi(x), \widetilde{V}_\ell[\xi(x)]\right\rangle\right\rangle. && (4.42)
\end{aligned}$$

We examine the terms (4.40)–(4.42) in turn. For (4.40), we have

$$\widetilde{V}_\ell[\xi(x)] \cdot \widetilde{V}_j[\xi(x)] = \delta_{j\,\ell}.$$

Since $\xi(x)$ is always in M, $D\xi(x)$ applied to *anything* must be a tangent vector. Thus, for (4.41), we have

$$\left\langle D\xi(x), \widetilde{V}_\ell[\xi(x)]\right\rangle \cdot \widetilde{V}_j[\xi(x)] = 0.$$

But even more is true. When we move in a normal direction, the nearest point does not change, so we have

$$\left\langle D\xi(x), \widetilde{V}_\ell[\xi(x)]\right\rangle = 0.$$

Thus we have

$$\left\langle D\widetilde{V}_j[\xi(x)], \left\langle D\xi(x), \widetilde{V}_\ell[\xi(x)]\right\rangle\right\rangle = 0,$$

so (4.42) vanishes.

Combining the values for (4.40)–(4.42), we obtain

$$\left\langle D\widetilde{Q}_j(x), \widetilde{V}_\ell[\xi(x)]\right\rangle = \delta_{j\,\ell}. \tag{4.43}$$

Next we consider any tangent direction T. We compute

$$\begin{aligned}
\left\langle D\widetilde{Q}_j(x), T[\xi(x)]\right\rangle &= \left\langle \mathbf{I} - D\xi(x), T[\xi(x)]\right\rangle \cdot \widetilde{V}_j[(\xi(x)] \\
&\quad + [x - \xi(x)] \cdot \left\langle D(\widetilde{V}_j \circ \xi)(x), T[(\xi(x)]\right\rangle \\
&= T[\xi(x)] \cdot \widetilde{V}_j[\xi(x)] && (4.44) \\
&\quad - \left\langle D\xi(x), T[\xi(x)]\right\rangle \cdot \widetilde{V}_j[\xi(x)] && (4.45) \\
&\quad + [x - \xi(x)] \cdot \left\langle D\widetilde{V}_j[\xi(x)], \left\langle D\xi(x), T[\xi(x)]\right\rangle\right\rangle. && (4.46)
\end{aligned}$$

[6] Recall from Section 3.3 that we use the notation $\langle\ ,\ \rangle$ to denote the application of the Jacobian matrix to a vector.

We see that the terms (4.44)–(4.46) vanish as follows:
(4.44) This term is trivially zero as it is the inner product of a unit tangent vector and a normal vector.
(4.45) Since $\xi = F \circ \Pi_1 \circ \Phi^{-1}$, the image of $D\xi$ must lie in the image of DF, that is, it must lie in the tangent space. Thus this term is also the inner product of a tangent vector and a normal vector.
(4.46) This term vanishes because $x - \xi(x)$ is a normal vector and

$$\left\langle D\widetilde{V}_j[\xi(x)], \left\langle D\xi(x), T[\xi(x)]\right\rangle\right\rangle$$

is a tangent vector, as guaranteed by Lemma 4.4.13(2).

Combining the values for (4.44)–(4.46), we obtain

$$\left\langle D\tilde{\varrho}_j(x), T[\xi(x)]\right\rangle = 0. \tag{4.47}$$

Let $T_1, T_2, \ldots, T_K$ be an orthonormal set of C^k tangent vector fields. From (4.43) and (4.47), we know how to represent $D\tilde{\varrho}$ in terms of the basis

$$\left\{T_1, T_2, \ldots, T_K, \widetilde{V}_1, \widetilde{V}_2, \ldots, \widetilde{V}_{N-K}\right\} \tag{4.48}$$

for $\mathbf{R}^N$. Specifically, in terms of the basis (4.48), we have

$$\begin{pmatrix} O_{K\times K} & O_{K\times(N-K)} \\ O_{(N-K)\times K} & I_{(N-K)\times(N-K)} \end{pmatrix}, \tag{4.49}$$

where O and I denote the zero and identity matrices of the indicated sizes. As the final step, we change basis in $\mathbf{R}^N$ from (4.48) to the standard basis. Since this change of basis is a C^{k-1} operation, we see that $D\tilde{\varrho}$ is C^{k-1}, and thus $\tilde{\varrho}$ is C^k. □

One may ask what happens in case M is only C^1. It no longer makes any sense to demand that the tangential derivatives of the normal vectors be tangent, because such derivatives need not exist. Nonetheless, when $K = N - 1$, the definition of $\tilde{\varrho}$ does not really require Lemma 4.4.13; in fact, Corollary 4.4.9 is sufficient. We have the following result

Theorem 4.4.17 *If $M \subset \mathbf{R}^N$ is a compact C^1 submanifold of dimension $N-1$ and if M has positive reach, then $\tilde{\varrho}$ is C^1 in a neighborhood of M.*

Proof. Let V be the unit normal field chosen so that

$$\tilde{\varrho}(x) = \left(x - \xi(x)\right) \cdot V[\xi(x)].$$

By Lemma 4.4.4 and Theorem 4.4.6, and the fact that

$$\tilde{\varrho}(x) = \left(x - \xi(x)\right) \cdot V[\xi(x)] = \operatorname{sign}\left[\left(x - \xi(x)\right) \cdot V[\xi(x)]\right] \varrho(x),$$

we see that, as long as x is in the interior of $\Xi(M) \setminus M$, then

$$\left\langle D\tilde{\varrho}(x), V[\xi(x)] \right\rangle = 1.$$

Similarly,

$$\left\langle D\tilde{\varrho}(x), T \right\rangle = 0.$$

holds for any direction that is tangent to M at $\xi(x)$. Thus

$$D\tilde{\varrho} = V[\xi(x)] \tag{4.50}$$

holds on the interior of $\Xi(M) \setminus M$. Repeating the argument used in the proof of Theorem 4.4.6, we see that (4.50) extends to the interior of $\Xi(M)$. Since (4.50) shows $D\tilde{\varrho}$ to be continuous, we conclude that $\tilde{\varrho}$ is C^1 in the interior of $\Xi(M)$, which is a neighborhood of M since M has positive reach. □

5

Variations and Generalizations

5.1 The Weierstrass Preparation Theorem

In Section 2.2, we described the method Newton devised for examining the local behavior of the locus of points satisfying a polynomial equation in two variables. It is easy to extend that method to the locus of an equation of the form

$$z^m + a_{m-1}(w)\,z^{m-1} + a_{m-2}(w)\,z^{m-2} + \cdots + a_0(w) = 0\,, \tag{5.1}$$

where each $a_i(w)$ is a holomorphic function of $w \in \mathbf{C}$ that vanishes at $w = 0$. Such an extension is significant, because the Weierstrass preparation theorem will show us that the behavior near $(0, 0)$ of the locus of an equation of the form given in (5.1) is completely representative of the local behavior of the locus of points satisfying $F(w, z) = 0$, where F is holomorphic.

Suppose $F(w, z)$, $(w, z) \in \mathbf{C}^n \times \mathbf{C}$, is holomorphic in a neighborhood of the origin, is not identically zero, and satisfies $F(0, 0) = 0$. To study the locus of the equation $F(w, z) = 0$ near the origin, we would apply the implicit function theorem if possible, but when the linear term in the Taylor series for F vanishes, the use of the implicit function theorem is *not* possible. Instead, the tool that can be used is the Weierstrass preparation theorem.

As a very simple example, consider the locus of points satisfying

$$\sin(z^2 - w) = 0 \tag{5.2}$$

near the origin. We can write

$$\sin(z^2 - w) = U(z^2 - w) \cdot (z^2 - w)$$

with $U(\xi) = \sin(\xi)/\xi$ (extended to equal 1 at the removable singularity $\xi = 0$). Since $U(\xi)$ is nonvanishing in a neighborhood of $\xi = 0$, there is a neighborhood of $(0, 0)$ in which $U(z^2 - w)$ is nonvanishing and, in that neighborhood, the locus of points in $\mathbf{C}^2$ satisfying (5.2) is identical to the locus of points satisfying

$$z^2 - w = 0\,.$$

This example worked out so easily because one could see in an obvious way how to factor out the nonvanishing function $U(z^2 - w)$. In the next example, the nonvanishing factor is less obvious. The more important aspect of the next example is that it illustrates a systematic procedure for constructing U. Following Hörmander [Ho 66], we will use that procedure for the main construction in the proof of the Weierstrass preparation theorem.

Example 5.1.1 Consider the locus of points satisfying

$$\frac{z^2}{1+z^2} + w = 0 \tag{5.3}$$

near the origin.

We will construct a function $S(w, z)$ that is nonvanishing and holomorphic in a neighborhood of the origin and that satisfies

$$z^2 + a_1(w)\,z + a_0(w) = \left(z^2 + w(1+z^2)\right) S(w, z)\,, \tag{5.4}$$

with $a_1(w)$ and $a_0(w)$ both holomorphic and with $a_1(0) = a_0(0) = 0$. Once those functions are constructed, we can set

$$U(w, z) = (1+z^2)^{-1}\, S(w, z)^{-1}$$

and multiply both sides of (5.4) by U to see that

$$U(w, z)\left(z^2 + a_1(w)\,z + a_0(w)\right) = \frac{z^2}{1+z^2} + w$$

holds. We conclude that, in a neighborhood of the origin, the locus of points satisfying (5.3) is identical to the locus of points satisfying

$$z^2 + a_1(w)\,z + a_0(w) = 0\,. \tag{5.5}$$

The functions in (5.4) will be constructed iteratively by setting

$$S_0 \equiv 1, \qquad R_0 \equiv 0$$

and, for $k = 1, 2, \ldots$, solving

$$z^2 - w(1+z^2)\, S_{k-1}(w, z) = z^2\, S_k(w, z) + R_k(w, z)\,, \tag{5.6}$$

where $R_k(w, z)$ is the remainder after the left-hand side of (5.6) is divided by z^2, that is, $R_k(w, z)$ is a linear expression in z with coefficients that are holomorphic in w and vanish at $w = 0$.

One computes

$$z^2 - w(1+z^2)\, S_0 = z^2 - w(1+z^2) = z^2(1-w) - w$$

so

$$S_1 = 1 - w, \qquad R_1 = -w\,.$$

Next, we compute

$$z^2 - w(1+z^2)\, S_1 = z^2 - w(1+z^2)(1-w) = z^2(1 - w + w^2) - w + w^2\,,$$

so

$$S_2 = 1 - w + w^2, \qquad R_2 = -w + w^2\,.$$

In general, one checks that

$$S_k(w, z) = \sum_{j=0}^{k} (-1)^j w^j\,, \qquad R_k(w) = \sum_{j=1}^{k} (-1)^j w^j\,.$$

Since S_k converges as $k \to \infty$ to $(1+w)^{-1}$ for $|w| < 1$, we conclude that

$$S(w, z) = S(w) = (1+w)^{-1}\,.$$

From (5.4) one can deduce that

$$a_1(w)\, z + a_0(w) = w/(1+w)\,.$$

(It is particular to this example that S depends only on w and that $a_1 \equiv 0$.) Alternatively, one can obtain the same result by using $R_k = \sum_{j=1}^{k} (-1)^j w^j$ and observing that

$$a_1(w)\, z + a_0(w) = -\lim_{k\to\infty} R_k(w) = w/(1+w)\,.$$

We see that, for $|w| < 1$ and $|z| < 1$, the locus of points satisfying (5.3) is identical to the locus of points satisfying

$$z^2 + \frac{w}{1+w} = 0\,.$$

□

The important feature of the polynomial $z^2 + a_1(w)z + a_0(w)$ in the example, as well as of the polynomial in (5.1), is that they are monic polynomials in z with holomorphic coefficients that vanish at $w = 0$. This class of polynomials is named in the next definition.

Definition 5.1.2 A function $W(w_1, w_2, \ldots, w_n, z)$, holomorphic in a neighborhood of $0 \in \mathbb{C}^{n+1}$, is called a *Weierstrass polynomial* of degree m, if, writing $w = (w_1, w_2, \ldots, w_n)$, we have

$$W(w,z) = z^m + a_{m-1}(w)\, z^{m-1} + \cdots + a_1(w)\, z + a_0(w)\,, \tag{5.7}$$

where each $a_i(w)$ is a holomorphic function in a neighborhood of $0 \in \mathbb{C}^n$ that vanishes at $w = 0 \in \mathbb{C}^n$.

Theorem 5.1.3 (Weierstrass Preparation Theorem) *Let $F(w,z)$, $w = (w_1, w_2, \ldots, w_n)$, be holomorphic in a neighborhood of $0 \in \mathbb{C}^{n+1}$. Let m be a positive integer. If $F(0,z)/z^m$ is holomorphic in a neighborhood of $0 \in \mathbb{C}$ and is non-zero at 0, then there exist a Weierstrass polynomial $W(w,z)$ of degree m and a function $U(w,z)$ holomorphic and nonvanishing in a neighborhood N of $0 \in \mathbb{C}^{n+1}$ such that*

$$F = UW \tag{5.8}$$

holds in N.

The proof of the Weierstrass Preparation Theorem 5.1.3 requires an estimate which we isolate in the following lemma.

Lemma 5.1.4 *Let m be a positive integer and let ρ_j, $j = 1, 2, \ldots, n+1$, be positive real numbers. Suppose*

$$z^m \phi(w,z) + \psi(w,z)$$

is holomorphic on

$$\Delta = \{(w,z) \;:\; |w_j| < \rho_j,\ j = 1, 2, \ldots, n,\ |z| < \rho_{n+1}\}$$

and suppose that ψ is a polynomial in z of degree less than m with coefficients that are holomorphic functions of w. If

$$B = \sup_{\Delta} |z^m \phi(w,z) + \psi(w,z)| < \infty\,,$$

then

$$\sup_{\Delta} |\psi(w,z)| \le mB \tag{5.9}$$

and

$$\sup_{\Delta} |\phi(w,z)| \le (m+1) B (\rho_{n+1})^{-m} \tag{5.10}$$

hold.

Proof of Lemma 5.1.4. Setting $f(w,z) = z^m\phi(w,z) + \psi(w,z)$, we have

$$\psi(w,z) = \sum_{j=0}^{m-1} \frac{\partial^j f(w,0)}{\partial z^j} z^j/j!. \tag{5.11}$$

Applying the Cauchy estimates (see Lemma 2.4.3) for each fixed w allows us to conclude that

$$\left|\frac{\partial^j f(w,0)}{\partial z^j}\right| \leq Bj!(\rho_{n+1})^{-j} \tag{5.12}$$

holds, for $j = 0, 1, \ldots, m-1$, so (5.9) follows from (5.11) and (5.12).

Now using (5.9) and selecting r with $0 < r < \rho_{n+1}$, for fixed w, we have

$$|\phi(w,z)| \leq |z^m\phi(w,z)|r^{-m} \leq (m+1)Br^{-m} \text{ for } |z| = r. \tag{5.13}$$

By the maximum modulus principle and the choice of r, (5.10) follows from (5.13). □

Proof of Theorem 5.1.3. Expressing F as a power series and collecting terms by powers of z, we can write

$$F(w,z) = \sum_{j=0}^{\infty} f_j(w)z^j.$$

Set

$$G(w,z) = \sum_{j=0}^{m-1} f_j(w)z^j$$

and

$$H(w,z) = z^{-m}\sum_{j=m}^{\infty} f_j(w)z^j,$$

so that

$$F = G + z^m H$$

and H is holomorphic and nonvanishing in a neighborhood of $0 \in \mathbb{C}^{n+1}$.

As in Example 5.1.1, we will construct the function S, holomorphic and nonvanishing in a neighborhood of $0 \in \mathbb{C}^{n+1}$ so that

$$\left(z^m + G/H\right)S \equiv W \tag{5.14}$$

is a Weierstrass polynomial. We can then define $U = H/S$ and the result will follow.

The iterative construction is defined by setting

$$S_0 \equiv 1, \qquad R_0 \equiv 0$$

and, for $k = 1, 2, \ldots$, by solving

$$z^m - (G/H)\, S_{k-1}(w, z) = z^m\, S_k(w, z) + R_k(w, z)\,, \tag{5.15}$$

where we require S_k to be holomorphic and nonvanishing in a neighborhood of $0 \in \mathbf{C}^{n+1}$ and we require $R_k(w, z)$ to be a polynomial in z of degree less than m with coefficients that are holomorphic functions of w.

Now we need to show that the sequence of functions $S_k(w, z)$, $k = 0, 1, \ldots$, is uniformly convergent in some neighborhood of $0 \in \mathbf{C}^{n+1}$. To establish this fact, we will need to apply Lemma 5.1.4. We choose the positive real numbers $\rho_1, \rho_2, \ldots, \rho_{n+1}$ so that

$$\sup_{\Delta} |G/H| \le \frac{(\rho_{n+1})^m}{2(m+1)}$$

holds, where

$$\Delta = \{(w, z) \;:\; |w_j| < \rho_j,\; j = 1, 2, \ldots, n,\; |z| < \rho_{n+1}\}\,.$$

We can make such a choice because $G(0, z) = 0$.

It follows from (5.15) that

$$\begin{aligned} z^m \Big(S_{k+1}(w, z) - S_k(w, z)\Big) &+ \Big(R_{k+1}(w, z) - R_k(w, z)\Big) \\ &= -(G/H)\Big(S_k(w, z) - S_{k-1}(w, z)\Big) \end{aligned} \tag{5.16}$$

holds. We conclude from Lemma 5.1.4 that

$$\sup_{\Delta} |S_{k+1}(w, z) - S_k(w, z)| \le \frac{1}{2} \sup_{\Delta} |S_k(w, z) - S_{k-1}(w, z)|\,.$$

Thus

$$S = \lim_{k \to \infty} S_k = \sum_{k=1}^{\infty} \Big(S_k - S_{k-1}\Big)$$

is uniformly convergent and, by (5.15), so is

$$R = \lim_{k \to \infty} R_k\,.$$

□

Remark 5.1.5

(1) An alternative, more elegant, but less elementary, proof of the Weierstrass preparation theorem can be found in D'Angelo [DA 93]. Bers [Be 64] gives proofs for both formal power series and convergent power series. An algebraic version of the Weierstrass preparation theorem proved by a similar iterative argument can be found in Gersten [Ge 83].

(2) For a function $F(w, z)$, $(w, z) \in \mathbf{C}^n \times \mathbf{C}$, that is holomorphic in a neighborhood of the origin, is not identically zero, and satisfies $F(0, 0) = 0$, it is always possible to make a linear change of variables so as to insure that the hypotheses of the Weierstrass Preparation Theorem 5.1.3 hold (exercise, or see Bers [Be 64]).

5.2 Implicit Function Theorems without Differentiability

In Section 3.4, we have already seen how to prove an implicit function theorem using the contraction mapping fixed point theorem. Here we will apply the more abstract Schauder fixed point theorem to obtain an implicit function theorem. The important practical distinction in the present approach is that we do not need differentiability.

We begin by stating the fixed point theorem of Schauder, which is a generalization of the Brouwer fixed point theorem in Euclidean space.

Theorem 5.2.1 (Schauder) *Let Y be a Banach space and let $K \subseteq Y$ be bounded, closed, and convex. If $F : K \to K$ is a compact operator,*[1] *then there exists a point $p \in K$ with $F(p) = p$.*

A proof of Schauder's theorem can be found in Zeidler [Ze 86]. The original reference is Schauder [Sc 30].

Note that there is no uniqueness of the fixed point guaranteed by Schauder's theorem. Our application to proving an implicit function theorem will also not provide a mechanism for obtaining a unique implicit value. But, by definition, a *function* must be single-valued. The dilemma then is to extract an ordinary single-valued function from a function that is naturally set-valued. The Axiom of Choice always provides one way to make such a selection of values, but that approach is far too abstract (and rarely leads to continuous or even measurable functions). Instead we will appeal to a measurable selection theorem. The result we use is the following (see Wagner [Wa 77; Theorem 4.1]).

Theorem 5.2.2 *If X is a topological space, Y is a complete, separable metric space, C_Y is the set of nonempty closed subsets of Y, and $\Phi : X \to C_Y$ is such that, for each open $U \subseteq Y$,*

$$X \cap \{x : \Phi(x) \cap Y \neq \emptyset\}$$

is a Borel set, then there exists a Borel measurable function $f : X \to Y$ with the property that

$$f(x) \in \Phi(x)$$

holds for each $x \in X$.

---

[1] That is, F is continuous and F maps bounded sets to relatively compact sets.

Remark 5.2.3 In fact, there is an entire literature of measurable selection theorems and, within that literature, a sub-topic of measurable implicit function theorems. The interested reader should consult Wagner [Wa 77] and the extensive bibliography therein.

We can now state and prove our theorem.

Theorem 5.2.4 *Let X be a topological space. Let Y be a separable Banach space and let $K \subseteq Y$ be compact and convex.*

If $F : X \times K \to K$ is continuous and if, for each $x \in X$, $F(x, \cdot)$ is a compact operator as a function of $y \in K$, then

(1) *for each $x \in X$,*
$$K \cap \{y : y = F(x, y)\}$$
is a nonempty closed set,

(2) *for each open $U \subseteq Y$,*
$$X \cap \left\{x : \emptyset \neq U \cap K \cap \{y : y = F(x, y)\}\right\}$$
is a Borel set, and

(3) *there is a Borel measurable function $f : X \to Y$ such that*
$$f(x) = F(x, f(x))$$
holds all for $x \in X$.

Proof.
(1) This is immediate from the continuity of F and Schauder's theorem.
(2) Because Y is separable and K is compact, we can write
$$U \cap K \cap \{y : y = F(x, y)\}$$
as a countable union of compact sets, that is,
$$U \cap K \cap \{y : y = F(x, y)\} = \bigcup_{i=1}^{\infty} C_i$$
with each C_i compact. We have
$$X \cap \left\{x : \emptyset \neq U \cap K \cap \{y : y = F(x, y)\}\right\} = \bigcup_{i=1}^{\infty} \Pi(C_i),$$
where $\Pi : X \times Y \to X$ is projection onto the first factor, and, for each i, $\Pi(C_i)$ is compact. Thus,
$$X \cap \left\{x : \emptyset \neq U \cap K \cap \{y : y = F(x, y)\}\right\}$$
is a Borel set.
(3) This conclusion follows from the measurable selection theorem (Theorem 5.2.2) above. □

5.3 An Inverse Function Theorem for Continuous Mappings

Examples show—see Section 3.6—that even a slight weakening of the hypothesis of C^1 smoothness in the classical implicit function theorem gives rise to a failure of the result. Mathematics is, nevertheless, robust. One can weaken the C^1 hypothesis and strengthen the properties of the mapping in other ways and still derive a suitable implicit function theorem. This is what we do in the present section. Our work here is inspired by that in Fečkan [Fe 94].

The device that we use to compensate for lack of smoothness is functional analysis, in particular, the mapping properties of nonlinear operators presented in Chapter 5 of Berger [Be 77]. Specifically, we will assume that our mapping is a compact[2] perturbation of the identity.

For the first part of this section, we let X be a reflexive Banach space. For instance X could be a Hilbert space or an L^p space with $1 < p < \infty$. Let $\overline{\mathbb{B}}$ be the closed unit ball in X. Consider a mapping $F : \overline{\mathbb{B}} \to X$ which is a continuous, compact perturbation of the identity: $F = I + G$, with G compact.

We will now generalize the concept of the "set-valued directional derivative" used in Kummer [Ku 91] to the context of Banach spaces. Let $\varphi : [0, \infty) \to [0, \infty)$ be a continuous mapping such that $\varphi(z) = 0$ if and only if $z = 0$. Let $\mathbb{S}$ be the set of limit points of $\partial\overline{\mathbb{B}}$. Here, and in all subsequent discussions, the limit in X is taken in the sense of the weak topology on X.

Definition 5.3.1 We define a *set-valued* directional derivative of the mapping F relative to φ at the point 0 in the direction $u \in \mathbb{S}$ to be

$$\begin{aligned}
&\Delta_\varphi F(0)u \\
&= \left\{ v : v = \lim_{k\to\infty} \frac{F(x_k + \lambda_k u_k) - F(x_k)}{\varphi(\lambda_k)}, \right. \\
&\qquad\qquad \left. \text{where } x_k \to 0,\ \lambda_k \downarrow 0,\ u_k \to u,\ \|u_k\| = 1 \right\}.
\end{aligned}$$

Remark 5.3.2 A similar definition was introduced by B. Kummer in [Ku 91] where he defined

$$\Delta F(0; u) = \left\{ v : v = \lim \frac{F(x_k + \lambda_k u) - F(x_k)}{\lambda_k},\ x_k \to 0,\ \lambda_k \downarrow 0 \right\}. \tag{5.17}$$

In case $\imath(z) = z$, X is finite dimensional, and F is locally Lipschitz, then we have

$$\|F(x_k + \lambda_k u_k) - F(x_k + \lambda_k u)\| \le L \cdot |\lambda_k| \cdot \|u_k - u\| \to 0 .$$

Therefore, in this situation, $\Delta_\imath F(0)u = \Delta F(0; u)$ holds, where $\Delta F(0; u)$ is as in (5.17). □

---

[2] A compact mapping sends bounded sets to relatively compact sets.

Our theorem is as follows:

Theorem 5.3.3 *If*

$$0 \notin \Delta_\varphi F(0)u \tag{5.18}$$

for each $u \in \mathbb{S}$, then F is locally invertible at 0.

Proof. We claim that (5.18) implies the existence of a number $\epsilon > 0$ such that

$$\|F(x_1) - F(x_2)\| \geq \epsilon\varphi(\|x_1 - x_2\|) \tag{5.19}$$

for each $x_1, x_2 \in \overline{\mathbb{B}}_\epsilon = \{x : \|x\| \leq \epsilon\}$. In fact, if (5.19) is not true, then there is a sequence $\{(x_{1,i}, x_{2,i})\}$ such that

$$\|F(x_{1,i}) - F(x_{2,i})\| \leq \frac{1}{i}\varphi(\|x_{1,i} - x_{2,i}\|),\ x_{1,i}, x_{2,i} \to 0,\ x_{1,i} \neq x_{2,i}. \tag{5.20}$$

We set

$$\begin{aligned} \lambda_i &= \|x_{2,i} - x_{1,i}\| \\ u_i &= \frac{x_{2,i} - x_{1,i}}{\|x_{2,i} - x_{1,i}\|} \\ x_i &= x_{1,i}\,. \end{aligned}$$

Then $x_{2,i} = x_i + \lambda_i u_i$. Since $\|u_i\| = 1$ and X is reflexive, we can assume that u_i converges to some $\overline{u} \in \mathbb{S}$ weakly (by the Eberlein–Šmuljan theorem—see Zeidler [Ze 86; page 777]). Then (5.20) implies that $0 \in \Delta_\varphi F(0)\overline{u}$, which is a contradiction to (5.18).

Thus we know that (5.19) holds. So F is one-to-one near 0. Since F is a continuous, compact perturbation of the identity, we can apply the invariance of domain theorem (5.4.11) of Berger [Be 77] to conclude that F is a local homeomorphism at 0. This completes the proof. □

Corollary 5.3.4 *If φ is increasing, then the inverse mapping $\psi = F^{-1}$ (given by the theorem) satisfies*

$$\|\psi(x_1) - \psi(x_2)\| \leq \varphi^{-1}\left(\frac{\|x_1 - x_2\|}{c}\right)$$

for some constant $c > 0$ and for x_1, x_2 near $F(0)$.

Proof. The result is immediate from (5.19). □

Next we give some examples to illustrate the ideas we have presented so far.

Example 5.3.5 Let $X = \mathbf{R}$ and $F(x) = x^{1/3}$. Of course we know by inspection that this function is invertible (globally). But it certainly does not satisfy the hypothesis of the standard inverse function theorem at 0. Let us see instead how it fits into the rubric of our new theorem.

Let $\mathfrak{i}(z) = z$. Clearly

$$\Delta_{\mathfrak{i}} F(0)u = \emptyset .$$

Thus, by our theorem, F is invertible at 0. Observe that this F is not even Lipschitz. □

In the next example, we use our theorem to show that an odd power of x defines an invertible function. Since even powers of x fail to be invertible near the origin, the argument we use must rely on some significant distinction between odd and even powers. We isolate the salient fact in the following lemma.

Lemma 5.3.6 *Let k be a positive integer. For $x, h \in \mathbf{R}$, it holds that*

$$|(x+h)^{2k} - x^{2k}| = 0 \quad \text{if and only if} \quad h = 0 \text{ or } x = -\tfrac{1}{2}h, \tag{5.21}$$

and

$$|(x+h)^{2k+1} - x^{2k+1}| \geq 2^{-2k} |h|^{2k+1} . \tag{5.22}$$

Proof. The result in (5.21) is clear. Likewise, (5.22) is clear if $h = 0$. Thus, we may assume that $h \neq 0$.

Using the binomial theorem, we compute

$$\begin{aligned} |(x+h)^{2k+1} - x^{2k+1}| &= \left| \left(\sum_{j=0}^{2k+1} \binom{2k+1}{j} x^{2k+1-j} h^j \right) - x^{2k+1} \right| \\ &= |h| \left| (2k+1)x^{2k} + \sum_{j=2}^{2k+1} \binom{2k+1}{j} x^{2k+1-j} h^{j-1} \right| . \end{aligned} \tag{5.23}$$

We see that the quantity in (5.23) clearly diverges to $+\infty$ as $|x| \to \infty$. Thus, the left-hand side of (5.22) must attain its minimum at some value of x. We find that value of x by differentiating $(x+h)^{2k+1} - x^{2k+1}$ (with respect to x) and setting that derivative equal to 0. That is, we need to solve

$$(x+h)^{2k} - x^{2k} = 0$$

for x. But, by (5.21) and the assumption that $h \neq 0$, we know that the only solution is when $x = -\frac{1}{2}h$. The result now follows by observing that, when $x = -\frac{1}{2}h$, (5.22) is an equality. □

Example 5.3.7 Let $X = \mathbf{R}$ and $F(x) = x^n$ for $n > 1$ an odd integer. Again, this F does not satisfy the hypothesis of the standard inverse function theorem at 0 (because the derivative at 0 vanishes). We set $\varphi(z) = z^n$. Since n is odd, we can use (5.22) to see that $v \in \Delta_\varphi F(0)u$ for $|u| = 1$ implies that $|v| \geq 2^{1-n}$. As a result, our theorem gives that F is invertible near 0.

Note by contrast that $0 \in \Delta F(0; u)$ (as in the remark following Definition 5.3.1). So more classical approaches to the inverse function theorem will *not* yield the local invertibility of F. □

For the next example we need to introduce some machinery. Namely, we consider the special case when $F(x) = x - G(x)$, for $G : X \to Y$, a locally Lipschitz mapping. We also assume that Y is a Banach space that is compactly imbedded into X. It turns out that this set of operators is well suited to the study of certain boundary value problems (as the next example will illustrate). In the present discussion, we no longer need assume that X is reflexive.

Theorem 5.3.8 *Let X, Y be Banach spaces with Y compactly imbedded into X. Consider $F(x) = x - G(x)$, where $G : X \to Y$ is locally Lipschitz. If $0 \notin \Delta F(0; u)$ for every u with $\|u\| = 1$, then F is locally invertible at 0.*

Remark 5.3.9 Note that it is immediate that $0 \in \Delta F(0; u)$ if and only if $u \in \Delta G(0; u)$.

Proof. Proceeding as in the proof of Theorem 5.3.3, we will show that $0 \notin \Delta F(0; u)$ implies the existence of a number $\epsilon > 0$ such that

$$\|F(x_1) - F(x_2)\|_Y \geq \epsilon \|x_1 - x_2\|_X \tag{5.24}$$

for each $x_1, x_2 \in \overline{\mathbb{B}}_{X,\epsilon} = X \cap \{x : \|x\| \leq \epsilon\}$. In fact, if (5.24) is not true, then there is a sequence $\{(x_{1,i}, x_{2,i})\}$ such that

$$\|F(x_{1,i}) - F(x_{2,i})\|_Y \leq \frac{1}{i}\left\|x_{1,i} - x_{2,i}\right\|_X, \quad x_{1,i}, x_{2,i} \to 0, \ x_{1,i} \neq x_{2,i}. \tag{5.25}$$

We set

$$\begin{aligned} \lambda_i &= \|x_{2,i} - x_{1,i}\|_X \\ u_i &= \frac{x_{2,i} - x_{1,i}}{\|x_{2,i} - x_{1,i}\|_X} \\ x_i &= x_{1,i} . \end{aligned}$$

Then $x_{2,i} = x_i + \lambda_i u_i$. By (5.24) and the fact that Y is compactly imbedded into X, we see that

$$\left\|u_i - \frac{G(x_i + \lambda_i) - G(x_i)}{\lambda_i}\right\|_Y = \frac{\|F(x_{2,i}) - F(x_{1,i})\|_Y}{\lambda_i} \leq \frac{1}{i}.$$

Since G is locally Lipschitz, we have

$$\left\|\frac{G(x_i + \lambda_i u_i) - G(x_i)}{\lambda_i}\right\|_Y \leq L\frac{\|\lambda_i u_i\|_X}{\lambda_i} = L .$$

Since Y is compactly imbedded into X, we see that we can pass to a subsequence, without changing notation, so that

$$\frac{G(x_i + \lambda_i u_i) - G(x_i)}{\lambda_i}$$

converges. Thus we also have that u_i converges to $u \in \mathbb{S}$.

Again using the fact that G is locally Lipschitz, we have

$$\|G(x_k + \lambda_k u_k) - G(x_k + \lambda_k u)\|_Y \leq L \cdot |\lambda_k| \cdot \|u_k - u\|_X \to 0 .$$

As a result, we have

$$\begin{aligned} 0 &= u - \lim_{i\to\infty} \frac{G(x_i + \lambda_i u) - G(x_i)}{\lambda_i} \\ &= \lim_{i\to\infty} \frac{F(x_i + \lambda_i u) - F(x_i)}{\lambda_i}, \end{aligned}$$

contradicting the assumption that $0 \notin \Delta F(0; u)$.

Thus we see that (5.24) holds, and the result now follows from the invariance of domain theorem (5.4.11) of Berger [Be 77]. □

Now we will introduce a new class of examples and ideas. For this we need a definition.

Definition 5.3.10 Consider a mapping $F : \mathbf{R}^m \to \mathbf{R}^n$. We say that F has the *property* $\mathcal{J}$ *at* 0 if there exist $r_0 > 0$, $\lambda_0 > 0$, a bounded set $\mathcal{M}_0$ of $n \times m$ matrices, and a constant $M_0 > 0$ such that, for each $x \in \mathbf{R}^m$ with $|x| \leq r_0$, each $z \in \mathbf{R}^m$ $|z| \leq 1$, and each $\lambda \in \mathbf{R}$ with $0 < \lambda \leq \lambda_0$, there is a matrix $A \in \mathcal{M}_0$ with

$$\left| \frac{F(x + \lambda z) - F(x)}{\lambda} - Az \right| \leq M_0(\lambda + |x|) .$$

For $u \in \mathbf{R}^m$, we will write

$$\Box_{\mathcal{M}_0} f(0; u) = \overline{\{Au : A \in \mathcal{M}_0\}} \subseteq \mathbf{R}^n . \tag{5.26}$$

Remark 5.3.11

(1) Observe that none of r_0, λ_0, $\mathcal{M}_0$, and M_0 in the definition is assumed to be unique, so of course, neither is $\Box_{\mathcal{M}_0} f(0; u)$.

(2) Note that $\Box_{\mathcal{M}_0} f(0; u)$ is compact for any choice of u.

(3) If f is C^2, then f has the property $\mathcal{J}$ at 0 and one may set $\mathcal{M}_0 = \{Df(0)\}$.

Example 5.3.12 Suppose $a \leq b$ are real numbers and $r : \mathbf{R} \to \mathbf{R}$ is C^2 with $r(0) = r'(0) = 0$. Set

$$f(t) = \begin{cases} r(t) + at & \text{for} \quad t \geq 0 \\ r(t) + bt & \text{for} \quad t \leq 0 . \end{cases}$$

Then the function $f(t)$ has the property $\mathcal{J}$ at 0 with

$$\mathcal{M} = \{A : At = ct, c \in [a, b]\} .$$

□

Example 5.3.13 Consider the boundary value problem (for an $\mathbf{R}^m$-valued function $u(t)$, $t \in \mathbf{R}$)

$$\begin{aligned} u'' &= f(u) + h(t) \\ u(0) &= u(1) = 0 \end{aligned} \tag{5.27}$$

where $f : \mathbf{R}^m \to \mathbf{R}^m$ is locally lipschitzian, has the property $\mathcal{J}$ at 0, and satisfies $f(0) = 0$. Also we assume that h is continuous.

Set

$$\begin{aligned} X &= C^0[0,1], \\ Y &= C^2[0,1], \\ F(x) &= x - G(x), \\ G(x) &= \mathcal{K}(f \circ x), \end{aligned}$$

where $\mathcal{K} : X \to Y$ is the inverse of the mapping from $C^2 \cap \{u : u(0) = u(1) = 0\}$ to C^0 given by $u \mapsto u''$. That is, if $x \in X = C^0[0,1]$, then $\mathcal{K}(x)$ is the unique C^2 function u (on $[0,1]$) with $u''(t) = x(t)$ and satisfying $u(0) = u(1) = 0$. Note that $\mathcal{K}$ is linear.

Preparatory to applying Theorem 5.3.8, we will investigate $\Delta G(0; u)$ for $u \in X$ with $\|u\| \equiv \sup\{|u(t)| : t \in [0,1]\} = 1$. Fix such a u and consider sequences $x_k \in X$ and $\lambda_k \in \mathbf{R}$, $k = 1, 2, \ldots$, with $\|x_k\| \le r_0$, $x_k \to 0$, $0 < \lambda_k \le \lambda_0$, and $\lambda_k \to 0$, where r_0 and λ_0 are as in Definition 5.3.10. For each k, we set

$$w_k = \frac{G(x_k + \lambda_k u) - G(x_k)}{\lambda_k} = \mathcal{K}\frac{f(x_k + \lambda_k u) - f(x_k)}{\lambda_k},$$

and suppose $w_k \to w \in Y$. For each k, we have

$$w_k = \mathcal{K} v_k,$$

where

$$v_k = \frac{f(x_k + \lambda_k u) - f(x_k)}{\lambda_k},$$

and because f has the property $\mathcal{J}$, we also have

$$\operatorname{dist}\Big[v_k(t), \Box_{\mathcal{M}_0} f(0; u(t))\Big] \le M_0(\lambda_k + |x_k(t)|) \text{ for } 0 \le t \le 1. \tag{5.28}$$

Setting

$$T(u) = \Big\{s \in X : s(t) \in \Box_{\mathcal{M}_0} f(0; u(t)) \text{ for } 0 \le t \le 1\Big\},$$

it follows from (5.28) that

$$w \in \overline{\mathcal{K}T(u)}.$$

We conclude that

$$\Delta G(0; u) \subseteq \overline{\mathcal{K}T(u)}.$$

As a consequence of Theorem 5.3.8, we have the following theorem:

If, for each $0 \neq u \in X = C^0[0,1]$, we have that $u \notin \overline{\mathcal{K}T(u)}$, then the system (5.27) has precisely one small (in the Banach space norm) solution for any continuous h with $\|h\|$ small.

5.4 Some Singular Cases of the Implicit Function Theorem

The standard implicit/inverse function theorem requires that the function in question be C^1 and that its Jacobian matrix be nondegenerate in a suitable sense, but simple examples show that something is still true even when the Jacobian matrix degenerates. For instance, the function $f(x) = x^3$ on the real line fails the Jacobian test at the origin. But $f''(0) = 0$ and $f'''(0) \neq 0$, and then elementary Taylor series considerations show that the function must be locally invertible at the origin. (Of course the function is obviously invertible just by inspection. Also the methods of Section 5.3, Example 5.3.7 could be applied, but we are now adopting the point of view of the inverse function theorem.)

The purpose of the present section is to explore the types of implicit and inverse function theorems that might be true in the case of a degenerate Jacobian matrix. Although there is an alternative treatment in Lefschetz [Le 57; pages 163–169], there does not seem to be any single all-encompassing theorem about this situation. We content ourselves here with the treatment of some illustrative special situations and some accompanying examples. We follow closely the treatment in Loud [Lo 61].

Preliminary Remarks

Let $F(x, y, z)$ and $G(x, y, z)$ be C^1 in a neighborhood of the origin and satisfy

$$F(0,0,0) = G(0,0,0) = 0.$$

Since the two equations

$$\begin{aligned} F(x,y,z) &= 0 \\ G(x,y,z) &= 0 \end{aligned} \tag{5.29}$$

involve three variables, the usual implicit function theorem paradigm is that two of the variables, say x and y, should be expressible as a function of the third variable, z. Of course, the hypothesis required to apply the implicit function theorem is the nonvanishing at the origin of the Jacobian

$$\mathcal{J} = \det \begin{pmatrix} F_x & F_y \\ G_x & G_y \end{pmatrix} \tag{5.30}$$

of the pair of functions F and G considered as a mapping

$$(x, y) \longmapsto \big(F(x,y,z), G(x,y,z)\big)$$

(so essentially z serves as a parameter).

In (5.30), and throughout this section, subscripts will denote partial derivatives. Additionally, we will use a superscript "o" to indicate that a partial derivative is to be evaluated at the origin, so for example

$$G_y^0 = \frac{\partial G}{\partial y}(0,0,0)$$

and later, when J is a function of two variables,

$$J_y^0 = \frac{\partial J}{\partial y}(0,0).$$

In this section we consider solving for x and y in terms of z in a neighborhood of $(0,0,0)$, even without the hypothesis that $\mathcal{J} \neq 0$. Of course, just as in the usual application of the implicit function theorem, if such functions $x(z)$ and $y(z)$ exist and are differentiable, then (by the chain rule) their derivatives dx/dz and dy/dz at $z = 0$ must satisfy

$$\begin{aligned} F_x \frac{dx}{dz} + F_y \frac{dy}{dz} + F_z &= 0, \\ G_x \frac{dx}{dz} + G_y \frac{dy}{dz} + G_z &= 0. \end{aligned}$$

Thus it is clearly necessary (for the existence and finiteness of these derivatives) that the matrix

$$\begin{pmatrix} F_x^0 & F_y^0 & F_z^0 \\ G_x^0 & G_y^0 & G_z^0 \end{pmatrix} \tag{5.31}$$

have the same rank as the Jacobian matrix

$$\begin{pmatrix} F_x^0 & F_y^0 \\ G_x^0 & G_y^0 \end{pmatrix}. \tag{5.32}$$

If in fact the matrix (5.31) has rank greater than the rank of the matrix (5.32), then the differentiable solution we seek does not exist.

The Case of Jacobian Matrix of Rank 1

We assume now that the rank of both the matrices (5.31) and (5.32) is 1. At least one of the entries in (5.32) is nonzero. For specificity, let us suppose that

$$\frac{\partial F}{\partial x}(0,0,0) = F_x^0 \neq 0.$$

Because (5.31) has rank 1 and because we have assumed $F_x^0 \neq 0$, it holds that

$$G_y^0 = \frac{G_x^0}{F_x^0} F_y^0 \quad \text{and} \quad G_z^0 = \frac{G_x^0}{F_x^0} F_z^0. \tag{5.33}$$

We will replace the system $F(x, y, z) = 0, G(x, y, z) = 0$ by an equivalent but simpler system. Set

$$H(x, y, z) = G(x, y, z) - \frac{G_x^0}{F_x^0} F(x, y, z).$$

It is obvious that $H_x^0 = 0$ and we can also apply (5.33), so we see that

$$H_x^0 = H_y^0 = H_z^0 = 0 \tag{5.34}$$

holds. Clearly, the system

$$\begin{aligned} F(x, y, z) &= 0 \\ H(x, y, z) &= 0 \end{aligned} \tag{5.35}$$

is equivalent to the original system (5.29).

Because we have made the normalizing assumption that $F_x^0 \neq 0$, we can apply the implicit function theorem to solve the first equation $F(x, y, z) = 0$ in (5.35) for x as a function of y and z near $y = z = 0$. If the result is written $x = f(y, z)$, then $f(0, 0) = 0$ and the partial derivatives of $f(y, z)$ can be computed from the partial derivatives of $F(x, y, z)$. In a neighborhood of $(0, 0)$, we have

$$f_y = -\frac{F_y[f(y, z), y, z]}{F_x[f(y, z), y, z]} \quad \text{and} \quad f_z = -\frac{F_z[f(y, z), y, z]}{F_x[f(y, z), y, z]}. \tag{5.36}$$

Substituting $x = f(y, z)$ into the second equation for $H(x, y, z)$ in (5.35), we eliminate the unknown x from the system.

We now consider the resulting equation

$$J(y, z) \equiv H[f(y, z), y, z] = 0. \tag{5.37}$$

If (5.37) is solved for y as a function of z, $y = y(z)$, with dy/dz finite at $z = 0$, then this $y(z)$, together with $x(z) = f(y(z), z)$, furnishes the desired solution of the system $F(x, y, z) = 0$, $H(x, y, z) = 0$. Furthermore, the derivative dx/dz would then be given by

$$\frac{dx}{dz} = -\frac{F_y \cdot (dy/dz) + F_z}{F_x}. \tag{5.38}$$

It remains to see how to solve for y as a function of z. If we assume that all the needed derivatives of F and H exist and are continuous, then the derivatives of $J(y, z)$ can be computed in terms of those of H and F by using (5.36). In particular, we see that

$$\begin{aligned} J_y &= (-H_x F_y + H_y F_x)/F_x, \\ J_z &= (-H_x F_z + H_z F_x)/F_x, \end{aligned}$$

so, by (5.34), we have $J_y^0 = J_z^0 = 0$.

The second and third derivatives of J are necessarily given by more complicated expressions. For instance, we have

$$\begin{aligned} J_{yy} &= (H_{xx}F_y^2 - 2H_{xy}F_yF_x + H_{yy}F_x^2)/F_x^2 \\ &\quad -H_x(F_y^2F_{xx} - 2F_xF_yF_{xy} + F_x^2F_{yy})/F_x^3 . \end{aligned}$$

While the second line in the preceding expression for J_{yy} is zero at $(0, 0)$ because $H_x^0 = 0$, there is no reason for the expression on the right in the first line to equal zero. More generally, the vanishing of J_{yy} or any other second or higher derivative of J at $(0, 0)$ would be a special occurrence indicating a relationship among the second or higher derivatives of F and G at the origin.

Since the first partial derivatives of $J(y, z)$ vanish at $(0, 0)$, we may write

$$\begin{aligned} J(y, z) &= \frac{1}{2}J_{yy}^0y^2 + J_{yz}^0yz + \frac{1}{2}J_{zz}^0z^2 + \frac{1}{6}J_{yyy}^0y^3 + \frac{1}{2}J_{yyz}^0y^2z \\ &\quad +\frac{1}{2}J_{yzz}^0yz^2 + \frac{1}{6}J_{zzz}^0z^3 + \text{higher-order terms} . \qquad (5.39) \end{aligned}$$

We want to find a solution of the equation $J(y, z) = 0$ which has $y = 0$ at $z = 0$ and for which z takes nonzero values. We write $y = \eta z$ in (5.39), where η will be a real-valued function of z. Then we find that

$$\begin{aligned} J(y, z) &= z^2\left(\frac{1}{2}J_{yy}^0\eta^2 + J_{yz}^0\eta + \frac{1}{2}J_{zz}^0\right) \\ &\quad +z^3\left(\frac{1}{6}J_{yyy}^0\eta^3 + \frac{1}{2}J_{yyz}^0\eta^2 + \frac{1}{2}J_{yzz}^0\eta + \frac{1}{6}J_{zzz}^0\right) \\ &\quad + \text{higher-order terms} , \qquad (5.40) \end{aligned}$$

so that for $z \neq 0$, $J(y, z) = 0$ is equivalent to

$$\begin{aligned} J(\eta, z) &= \left(\frac{1}{2}J_{yy}^0\eta^2 + J_{yz}^0\eta + \frac{1}{2}J_{zz}^0\right) \\ &\quad z\left(\frac{1}{6}J_{yyy}^0\eta^3 + \frac{1}{2}J_{yyz}^0\eta^2 + \frac{1}{2}J_{yzz}^0\eta + \frac{1}{6}J_{zzz}^0\right) \\ &\quad + \text{higher-order terms} = 0 . \qquad (5.41) \end{aligned}$$

We shall solve (5.41) for η as a function of z. The solution of $J(y, z) = 0$ for the variable y will then by given by $y(z) = z\eta(z)$.

Consider the quadratic equation

$$\frac{1}{2}J_{yy}^0\eta^2 + J_{yz}^0\eta + \frac{1}{2}J_{zz}^0 = 0 , \qquad (5.42)$$

which is just the limit of the equation (5.41) as $z \to 0$. If $\eta = \eta(z)$ is a solution of (5.41), then $\eta(0)$ must satisfy (5.42). Now there are four cases depending on the sign of the discriminant $(J_{yz}^0)^2 - J_{yy}^0J_{zz}^0$ and the signs of the various second partial derivatives J_{yy}^0, J_{yz}^0, and J_{zz}^0. Typically, either Case I or Case II below would apply.

Case I:

$$(J_{yz}^0)^2 - J_{yy}^0 J_{zz}^0 < 0$$

or

$$J_{yy}^0 = J_{yz}^0 = 0 \textbf{ and } J_{zz}^0 \neq 0$$

In this case, (5.42) has no real roots. Thus no real value for $\eta(0)$ can exist, so a solution of the type we seek does not exist.

Case II:

$$(J_{yz}^0)^2 - J_{yy}^0 J_{zz}^0 > 0$$

In this case, (5.42) has one or two simple real roots, according as $J_{yy}^0 = 0$ or $J_{yy}^0 \neq 0$. In either case, the ordinary implicit function theorem then guarantees a solution $\eta = \eta(z)$ of the equation (5.41) for each such simple real root of (5.42). If η_0 is such a real root, then we have

$$\begin{aligned}
\eta(z) &= \eta_0 + o(1)\,, \\
y(z) &= z\eta(z) = \eta_0(z) + o(z)\,, \\
\frac{dy}{dz} &= \eta_0 \text{ at } z = 0\,.
\end{aligned}$$

From (5.38) we find that at $z = 0$ we have

$$\frac{dx}{dz} = -\frac{F_y\eta_0 + F_z}{F_x}\,. \tag{5.43}$$

Case III:

$$(J_{yz}^0)^2 - J_{yy}^0 J_{zz}^0 = 0 \textbf{ and } J_{yy}^0 \neq 0$$

In this case, we can still solve for $\eta(z)$ either for positive z or (in certain cases) for negative z. However, we will have to utilize fractional powers of z (very much in the spirit of Puiseux series—see Krantz and Parks [KP 92]). Let η_0 be a double root of (5.42). The equation (5.41) then becomes

$$\begin{aligned}
\frac{1}{2}J_{yy}^0(\eta-\eta_0)^2 + z\biggl(\frac{1}{6}J_{yyy}^0(\eta-\eta_0)^3 &+ J_1(\eta-\eta_0)^2 + J_2(\eta-\eta_0) + J_3\biggr) \\
&+ \text{higher-order terms} = 0\,.
\end{aligned}$$

Here

$$\begin{aligned}
J_1 &= \frac{1}{2}J_{yyy}^0\eta_0 + \frac{1}{2}J_{yyz}^0\,, \\
J_2 &= \frac{1}{2}J_{yyy}^0\eta_0^2 + J_{yyz}^0\eta_0 + \frac{1}{2}J_{yzz}^0\,, \\
J_3 &= \frac{1}{6}J_{yyy}^0\eta_0^3 + \frac{1}{2}J_{yyz}^0\eta_0^2 + \frac{1}{2}J_{yzz}^0\eta_0 + \frac{1}{6}J_{zzz}^0\,.
\end{aligned}$$

Subcase III(a):

$$J_3 = 0$$

This subcase cannot be resolved without the use of even higher order derivatives, so we will say no more about it.

Subcase III(b):

$$J_3 \neq 0$$

We replace z by u^2 if J_3 and J^0_{yy} have opposite signs; we replace z by $-u^2$ if J_3 and J^0_{yy} have the same sign. Then, taking a square root, we find that either

$$\eta - \eta_0 = \pm\sqrt{(-2J_3/J^0_{yy})u} + \text{ higher-order terms}$$

or

$$\eta - \eta_0 = \pm\sqrt{(2J_3/J^0_{yy})u} + \text{ higher-order terms}\,.$$

As a result, we have these solutions:

> If J_3 and J^0_{yy} have opposite signs, then there are two solutions $\eta(z)$ for positive z and none for negative z.
>
> If case J_3 and J^0_{yy} have the same sign, then there are two solutions $\eta(z)$ for negative z and none for positive z.

In either case, we have $y = \eta_0 z + O(|z|^{3/2})$ so that, at $z = 0$, $dy/dz = \eta_0$ and dx/dz is given by (5.43).

Case IV:

$$J^0_{yy} = J^0_{yz} = J^0_{zz} = 0$$

In this case, we can replace (5.40) by

$$\frac{1}{6}J^0_{yyy}\eta^3 + \frac{1}{2}J^0_{yyz}\eta^2 + \frac{1}{2}J^0_{yzz}\eta + \frac{1}{6}J^0_{zzz} + \text{ higher-order terms } = 0\,. \tag{5.44}$$

Consider now the equation

$$\frac{1}{6}J^0_{yyy}\eta^3 + \frac{1}{2}J^0_{yyz}\eta^2 + \frac{1}{2}J^0_{yzz}\eta + \frac{1}{6}J^0_{zzz} = 0\,, \tag{5.45}$$

which is the limit of (5.44) as $z \to 0$. If (5.45) has no real roots, then there are no solutions of the type that we seek. If η_0 is a multiple real root of (5.45), then higher derivatives are needed to produce a solution and we will say nothing about it here. Finally, if η_0 is a simple real root of (5.45), then the ordinary implicit function theorem shows that (5.44) has a solution $\eta = \eta(z)$ with $\eta(z) = \eta_0 + o(1)$. So, again, $y(z) = \eta_0 z + o(z)$; and, at $z = 0$,

$$\frac{dy}{dz} = \eta_0 , \qquad \frac{dx}{dz} = -\frac{F_y \eta_0 + F_z}{F_x} .$$

In this final case, there may be as many as three different solutions $x = x(z)$, $y = y(z)$.

The Case of Jacobian Matrix of Rank 0

In fact this situation is even more complicated, and more technical, than the rank 1 case. We shall not treat it in any detail, but instead content ourselves with presenting some examples. The interested reader is referred to Loud [Lo 61] for a more complete picture.

Example 5.4.1 Consider the system

$$\begin{array}{ccccccccccc} xz & + & yz & + & xy & - & z^3 & + & z^2 x & = & 0, \\ xz & + & 2yz & - & xy & - & 3z^3 & - & z^2 y & = & 0. \end{array} \tag{5.46}$$

Since there are no linear terms in the system (5.46), it is clear that the rank of the Jacobian matrix is zero at the origin. Nonetheless, we wish to solve (5.46) for x and y as functions of z. This can be done by eliminating a variable algebraically, but we will illustrate an approach that uses the implicit function theorem and thus can be applied more generally. Our method exploits the fact that the coefficient of z^2 is 0 in both equations.

Set $x = \xi z$ and $y = \eta z$. After eliminating common factors of z^2, we obtain the system

$$\begin{array}{ccccccccccc} \xi & + & \eta & - & z & + & z\xi & + & \xi\eta & = & 0, \\ \xi & + & 2\eta & - & 3z & - & z\eta & - & \xi\eta & = & 0. \end{array} \tag{5.47}$$

The pair of functions on the left-hand side of (5.47),

$$\begin{aligned} F(\xi, \eta, z) &= \xi + \eta - z + z\xi, \\ G(\xi, \eta, z) &= \xi + 2\eta - 3z - z\eta, \end{aligned}$$

satisfies

$$\begin{pmatrix} F_\xi^0 & F_\eta^0 \\ G_\xi^0 & G_\eta^0 \end{pmatrix} = \begin{pmatrix} 1 & 1 \\ 1 & 2 \end{pmatrix}, \tag{5.48}$$

and obviously $(\xi, \eta, z) = (0, 0, 0)$ is a point that satisfies (5.47). Since the matrix in (5.48) has rank 2, the implicit function theorem guarantees the existence of solutions $\xi(z)$ and $\eta(z)$ to the system (5.47) with $\xi(0) = \eta(0) = 0$. Also, the implicit function theorem allows us to compute the derivatives of $\xi(z)$ and $\eta(z)$ implicitly. We find that $d\xi/dz(0) = -1$ and $d\eta/dz(0) = 2$. Thus, we have

$$x(z) = -z^2 + o(z^2) , \qquad y(z) = 2z^2 + o(z^2) .$$

Finally, by inspection, one can see that all points in

$$\{(x, 0, 0) : x \in \mathbf{R}\} \cup \{(0, y, 0) : y \in \mathbf{R}\}$$

also solve (5.46), giving two additional solution curves passing through the origin. □

Remark 5.4.2 Example 5.4.1 is representative of what often happens with a pair of equations

$$\begin{aligned} F(x, y, z) &= 0, \\ G(x, y, z) &= 0. \end{aligned} \tag{5.49}$$

in which, for both, the nontrivial part of the Taylor polynomial begins with the quadratic term. Letting H_F and H_G be those nontrivial quadratic forms, we consider the system

$$\begin{aligned} H_F(x, y, z) &= 0, \\ H_G(x, y, z) &= 0, \\ x^2 + y^2 + z^2 &= 1. \end{aligned} \tag{5.50}$$

If a unit vector, $\mathbf{v}$, can be found that solves the system (5.50) and if the matrix of partial derivatives on the left-hand side of (5.50) is non-singular at $\mathbf{v}$, then one can form a new orthonormal basis for $\mathbf{R}^3$ which contains the vector $\mathbf{v}$, and after changing basis, the problem of solving (5.49) will look like Example 5.2.1 in that the square of one of the new variables will have coefficient 0. □

The next example illustrates how one may still be able to analyze the solutions of a pair of equations even when the procedure discussed in Remark 5.4.2 is not fruitful.

Example 5.4.3 Consider the system

$$\begin{aligned} x^2 + y^2 - z^2 + z^2x &= 0, \\ x^2 + 2y^2 - z^2 + z^2y &= 0. \end{aligned} \tag{5.51}$$

Again there are no linear terms, and thus the rank of the Jacobian matrix is zero at the origin.

Remark 5.4.2 suggests the change of variables $x = (1/\sqrt{2})u - (1/\sqrt{2})w$, $y = v$, $z = (1/\sqrt{2})u + (1/\sqrt{2})w$, which results in equations with neither the u^2 term nor the w^2 term. Unfortunately, when we replace v by ξu and w by ηu (or alternatively, replace u by ξw and v by ηw) and eliminate the common factors from each equation, we do not obtain a system with full rank (that is, rank 2) Jacobian matrix with respect to ξ and η.

Instead, to solve (5.51) for x and y as functions of z, we set $x = \xi z$ and $y = \eta z$. The result is the system

$$\begin{aligned} \xi^2 + \eta^2 - 1 + z\xi &= 0, \\ \xi^2 + 2\eta^2 - 1 + z\eta &= 0. \end{aligned} \tag{5.52}$$

Considering the limit of the system (5.52) as z approaches 0, we are led to the (real) common points of the loci of the equations

$$\xi^2 + \eta^2 - 1 = 0 \qquad \text{and} \qquad \xi^2 + 2\eta^2 - 1 = 0$$

which are $(1, 0)$ and $(-1, 0)$. At (1,0), we set $u = \xi - 1$, $v = \eta$. The system (5.52) becomes

$$\begin{array}{ccccccccc} 2u & + & u^2 & + & v^2 & + & z(u+1) & = & 0\,, \\ 2u & + & u^2 & + & 2v^2 & + & zv & = & 0\,. \end{array} \tag{5.53}$$

If we define the two functions

$$\begin{aligned} F(u, v, z) &= 2u + u^2 + v^2 + z(u+1) \\ G(u, v, z) &= 2u + u^2 + 2v^2 + zv\,, \end{aligned}$$

then we see that

$$\begin{pmatrix} F_u^0 & F_v^0 \\ G_u^0 & G_v^0 \end{pmatrix} = \begin{pmatrix} 2 & 0 \\ 2 & 0 \end{pmatrix} \tag{5.54}$$

is only of rank 1, so the implicit function theorem cannot be invoked to give us functions $u(z)$ and $v(z)$. On the other hand, it is the case that the matrix

$$\begin{pmatrix} F_u^0 & F_v^0 & F_z^0 \\ G_u^0 & G_v^0 & G_z^0 \end{pmatrix} = \begin{pmatrix} 2 & 0 & 1 \\ 2 & 0 & 0 \end{pmatrix} \tag{5.55}$$

has rank 2, so we will still be able to solve for u and v by elimination. The first equation of (5.53), when solved for u, gives

$$u = -\frac{1}{2}z - \frac{1}{2}v^2 + \frac{1}{8}z^2 + \text{ higher-order terms}\,.$$

When we substitute this last equation into the second equation of (5.53) we obtain

$$-z + v^2 + vz + \frac{1}{2}z^2 + \text{higher-order terms } = 0\,.$$

From this equation we see that, for small positive z, we can write

$$v = \pm\sqrt{z} + o(\sqrt{z})\,.$$

In conclusion, we find that

$$u = o(\sqrt{z})\,, \quad \xi = 1 = o(\sqrt{z})\,, \quad \eta = \pm\sqrt{z} + o(\sqrt{z})\,,$$

$$x = z + o(z^{3/2})\,, \quad y = \pm z^{3/2} + o(z^{3/2})\,,$$

all for small, positive z. The analysis at the point $\xi = -1$, $\eta = 0$ is similar. □

6

Advanced Implicit Function Theorems

6.1 Analytic Implicit Function Theorems

We will now consider implicit function theorems in both the real analytic and the complex analytic (holomorphic) categories. These are obviously closely related, as the problem in the real analytic category can be complexified (by replacing every x^j with a z^j) and thereby turned into a holomorphic problem. Conversely, any complex analytic implicit function theorem situation is *a fortiori* real analytic and can therefore be treated with real analytic techniques. And both categories are subcategories of the C^∞ category.

To make things completely clear we repeat: An implicit function problem with real analytic data automatically has a C^∞ solution by the classical C^∞ implicit function theorem. The point is to see that it has a real analytic solution. Likewise, an implicit function problem with complex analytic (holomorphic) data automatically has a C^∞ solution by the classical C^∞ implicit function theorem; it also automatically has a real analytic solution by the real analytic implicit function theorem. The point is to see that it has a *complex analytic* (holomorphic) solution.

The astute reader will notice that, in the theorem below, we use Cauchy's method of majorization. This is also the principal tool in the celebrated Cauchy–Kowalewsky theorem for existence and uniqueness of solutions to partial differential equations with real analytic data (see Section 4.1). In fact the implicit function theorem can be proved using a suitable existence and uniqueness theorem for partial differential equations, as was shown in Section 4.1.

In the following theorem and its proof we shall use *multiindex* notation. On $\mathbf{R}^N$, a multiindex is an N-tuple $\alpha = (\alpha_1, \ldots, \alpha_N)$, where each α_j is a nonnegative integer. Then, for $x \in \mathbf{R}^N$,

$$x^\alpha \equiv x_1^{\alpha_1} \cdot x_2^{\alpha_2} \cdots x_N^{\alpha_N} .$$

Also $|\alpha| = \alpha_1 + \cdots + \alpha_N$ and $\alpha! = \alpha_1! \cdot \alpha_2! \cdots \alpha_N!$. Finally, if $\alpha = (\alpha_1, \ldots, \alpha_N)$ and $\beta = (\beta_1, \ldots, \beta_N)$ are both multiindices, then $\alpha + \beta \equiv (\alpha_1 + \beta_1, \ldots, \alpha_N + \beta_N)$. We will use the notation $\mathbf{e}_j$ to denote the multiindex $(0, 0, \ldots, 1, 0, \ldots, 0)$, with a 1 in the jth position and all other entries 0. In calculations involving multiple variables, multiindex notation is essential for clarity.

We will consider power series expansions for a function $\varphi(x, y)$, where $x \in \mathbf{R}^N$ and $y \in \mathbf{R}$. So we will have powers x^α, for α a multiindex, and y^k, for k a nonnegative integer. We shall write such a power series expansion as

$$\varphi(x, y) = \sum_{\alpha,k} a_{\alpha,k} x^\alpha y^k .$$

It will be understood in this context that α ranges over all multiindices and k ranges from 0 to ∞. If we write $a_{0,0}$, it will therefore be understood that the first 0 is an N-tuple $(0, \ldots, 0)$ and the second is the single digit 0.

Finally, we need Hadamard's estimates for the size of the coefficients of a convergent power series that are given in the next lemma.

Lemma 6.1.1 *Let*

$$F(x) = \sum_{\alpha,k} a_\alpha x^\alpha y^k$$

be a function defined by a power series in $x = (x_1, \ldots, x_N)$ and y that is convergent for $|x| < R_1$, $|y| < R_2$. Assume that F is bounded by K. Then

$$|a_{\alpha,k}| \leq K \cdot R_1^{-|\alpha|} \cdot R_2^{-k} .$$

Proof. Exercise for the reader. Note that the constant K comes from an application of the root test. □

Theorem 6.1.2 *Suppose that the power series*

$$F(x, y) = \sum_{\alpha,k} a_{\alpha,k} x^\alpha y^k \tag{6.1}$$

is absolutely convergent for $|x| \leq R_1$, $|y| \leq R_2$. If

$$a_{0,0} = 0 \text{ and } a_{0,1} \neq 0, \tag{6.2}$$

then there exist $r_0 > 0$ and a power series

$$f(x) = \sum_{|\alpha|>0} c_\alpha x^\alpha \tag{6.3}$$

such that (6.3) is absolutely convergent for $|x| \leq r_0$ and

$$F(x, f(x)) = 0. \tag{6.4}$$

Proof. It will be no loss of generality to assume $a_{0,1} = 1$, so that (6.1) takes the form

$$F(x, y) = y + \sum_{|\alpha|>0} (a_{\alpha,0} + a_{\alpha,1} y) x^\alpha + \sum_{|\alpha|\geq 0} \sum_{k=2}^{\infty} a_{\alpha,k} x^\alpha y^k. \tag{6.5}$$

Introducing the notation $b_{\alpha,k} = -a_{\alpha,k}$, we can rewrite the equation $F(x, y) = 0$ as

$$y = \sum_{|\alpha|>0} (b_{\alpha,0} + b_{\alpha,1} y) x^\alpha + \sum_{|\alpha|\geq 0} \sum_{k=2}^{\infty} b_{\alpha,k} x^\alpha y^k, \tag{6.6}$$

or $y = B(x, y)$, where

$$B(x, y) = \sum_{|\alpha|>0} (b_{\alpha,0} + b_{\alpha,1} y) x^\alpha + \sum_{|\alpha|\geq 0} \sum_{k=2}^{\infty} b_{\alpha,k} x^\alpha y^k. \tag{6.7}$$

Substituting $y = f(x)$ into (6.6) with $f(x)$ given by (6.3), we obtain

$$\begin{aligned} \sum_{|\alpha|>0} c_\alpha x^\alpha &= \sum_{|\alpha|>0} b_{\alpha,0} x^\alpha + \sum_{|\alpha|>0} \sum_{|\beta|>0} b_{\alpha,1} c_\beta x^{\alpha+\beta} \\ &\quad + \sum_{|\alpha|\geq 0} \sum_{k=2}^{\infty} b_{\alpha,k} x^\alpha \left(\sum_{|\beta|>0} c_\beta x^\beta \right)^k . \end{aligned} \tag{6.8}$$

If all the series in (6.8) are ultimately shown to be absolutely convergent, then the order of summation can be freely rearranged. Assuming absolute convergence, we can equate like powers of x on the left-hand and right-hand sides of (6.8) and obtain the following recurrence relations: First, we have

$$c_{\mathbf{e}_j} = b_{\mathbf{e}_j,0}, \tag{6.9}$$

Equation (6.9) allows us to solve for each $c_{\mathbf{e}_j}$. Next we indicate how each coefficient c_α of higher index may be expressed in terms of the $b_{\gamma,j}$ and indices c_β with index of lower order. In point of fact, let us assume inductively that we have so solved for c_α for all multiindices α with $|\alpha| \leq p$. Then, fixing a multiindex α with $|\alpha| = p$ and identifying like powers of x, we find that

$$\begin{aligned} c_{\alpha+\mathbf{e}_j} &= b_{\alpha+\mathbf{e}_j,0} + \sum_{\substack{|\beta|>0, |\gamma|>0, \\ \beta+\gamma=\alpha+\mathbf{e}_j}} b_{\gamma,1} c_\beta \\ &\quad + \sum_{\substack{|\gamma|\geq 0, k\geq 2, \\ |\beta^\ell|>0, \beta^1+\cdots\beta^k+\gamma=\alpha+\mathbf{e}_j}} M(\beta^1, \ldots, \beta^k) \cdot b_{\gamma,k} \cdot c_{\beta^1} \cdot c_{\beta^2} \cdots c_{\beta^k} . \end{aligned} \tag{6.10}$$

Here $M(\beta^1, \ldots, \beta^k)$ is a suitable multinomial coefficient (and superscripts are labels, not powers). Observe that each such M is positive. We can see by inspection that all of multiindices on coefficients c that occur on the right-hand side have size less than or equal to p. This is the desired recursion.

While the recurrence relations (6.9) and (6.10) uniquely determine the coefficients c_β in the power series for the implicit function, it is also necessary to show that (6.3) is convergent. The easiest way to obtain the needed estimates is by using the method of majorants, which we describe next.

Consider two power series in the same number of variables:

$$\Phi(x_1, x_2, \ldots, x_\rho) = \sum_{j_1, j_2, \ldots, j_\rho = 0}^{\infty} \phi_{j_1, j_2, \ldots, j_\rho} x_1^{j_1} x_2^{j_2} \ldots x_\rho^{j_\rho}, \tag{6.11}$$

$$\Psi(x_1, x_2, \ldots, x_\rho) = \sum_{j_1, j_2, \ldots, j_\rho = 0}^{\infty} \psi_{j_1, j_2, \ldots, j_\rho} x_1^{j_1} x_2^{j_2} \ldots x_\rho^{j_\rho} \tag{6.12}$$

We say that $\Psi(x_1, x_2, \ldots, x_\rho)$ is a *majorant* of $\Phi(x_1, x_2, \ldots, x_\rho)$ if

$$|\phi_{j_1, j_2, \ldots, j_\rho}| \leq \psi_{j_1, j_2, \ldots, j_\rho} \tag{6.13}$$

holds for all $j_1, j_2, \ldots, j_\rho$.

Because all the coefficients $M(\beta^1, \ldots, \beta^k)$ in (6.10) are positive, it follows that if

$$G(x, y) = \sum_{|\alpha| \geq 0, k \geq 0} g_{\alpha,k} x^\alpha y^k,$$

(with $g_{0,0} = g_{0,1} = 0$) is a majorant of

$$B(x, y) = \sum_{|\alpha|>0} (b_{\alpha,0} + b_{\alpha,1} y) x^\alpha + \sum_{|\alpha| \geq 0, k \geq 2} b_{\alpha,k} x^\alpha y^k$$

and if

$$h(x) = \sum_{|\alpha| \geq 1} h_\alpha x^\alpha \tag{6.14}$$

solves

$$h(x) = G[x, h(x)], \tag{6.15}$$

then $h(x)$ will be a majorant of $f(x)$. Consequently, if the series (6.14) for $h(x)$ is convergent, then the series (6.3) is convergent and its radius of convergence is at least as large as the radius of convergence for (6.14).

We take

$$G(x, y) = \frac{Kr}{r - (x_1 + \cdots + x_N) - y},$$

where

$$K = \sup\{|B(x, y)| : x \in \overline{D}(0, R_1), \ y \in \overline{D}(0, R_2)\}$$

and r is sufficiently small (depending on R_1 and R_2). We see that G is a majorant of B by Lemma 6.1.1. For this choice of majorant, the equation (6.15) is quadratic and can be solved explicitly. The solution is clearly holomorphic at $x = 0$. In fact, $y = h(x)$ is easily seen to be the solution of the quadratic equation

$$y^2 + (x_1 + \cdots x_N - r)y + Kr = 0. \tag{6.16}$$

□

Theorem 6.1.2, the main result of this section, is somewhat special. For it only considers the case of one dependent variable and arbitrarily many independent variables. We leave it to the reader to apply the inductive method of Dini (Section 3.2) to derive a general real analytic implicit function theorem from our Theorem 6.1.2.

Recall that we proved a version of Theorem 6.1.2 in Section 2.4 during our discussion of Cauchy's contributions to this subject. We refer the reader to [Kr 92] and [KP 92] for a detailed consideration of various kinds of analytic implicit function theorems.

We close this section by noting that the complex analytic version of the implicit function theorem follows almost immediately from the C^1 version. That is, once you know that the solution $y = \psi(x)$ to a holomorphic system

$$F(x, y) = 0$$

is continuously differentiable, then you can apply $\partial/\partial\bar{z}_j$ to both sides and apply the chain rule to determine that ψ is holomorphic. Once the complex analytic case is proved, then the real analytic case follows easily (by the method of complexification) as already indicated. Further details of these ideas may be found in [Kr 92].

6.2 Hadamard's Global Inverse Function Theorem

We have already established that the implicit function theorem and the inverse function theorem are equivalent, so it is sometimes useful to refer to the two theorems interchangeably. In this section, we will formulate the relevant ideas in terms of the inverse function theorem.

Imagine a continuously differentiable mapping $F : U \to V$, where U, V are open subsets of $\mathbf{R}^N$. Assume that the Jacobian determinant of F is nonzero at every point. Then we may be sure that F is locally one-to-one. We now wish to consider under what circumstances F is *globally* one-to-one. This is not expected to hold in every case. For example, the mapping $(x, y) \mapsto (e^x \cos y,\ e^x \sin y)$ is smooth, locally one-to-one, and its Jacobian determinant is everywhere nonzero, but nevertheless the mapping is not globally one-to-one.

Let $F : \mathbf{R}^N \to \mathbf{R}^N$ be a C^2 mapping. For convenience of notation, we will assume $F(0) = 0$. We define $H : \mathbf{R} \times \mathbf{R}^N \to \mathbf{R}^N$ by setting

$$H(t, x) = \begin{cases} F(tx)/t & \text{if } 0 < t \\ \langle DF(0), x \rangle & \text{if } t \leq 0 . \end{cases}$$

(Recall from Section 3.3 that we use the notation $\langle\ ,\ \rangle$ to denote the application of the Jacobian matrix to a vector.) Note that $H(0, \cdot)$ is the Jacobian matrix of F, while $H(1, \cdot)$ is F, so H restricted to $[0, 1] \times \mathbf{R}^N$ is a homotopy between F and its Jacobian. It is not difficult to see that H is C^1 when F is C^2. In fact, we have

$$\frac{\partial H}{\partial x_i}(t, x) = \begin{cases} \dfrac{\partial F}{\partial x_i}(tx) & \text{if } 0 < t \\ \dfrac{\partial F}{\partial x_i}(0) & \text{if } t \leq 0 . \end{cases} \tag{6.17}$$

and

$$\frac{\partial H}{\partial t}(t, x) = \begin{cases} \langle DF(tx), x \rangle - F(tx)/t^2 & \text{if } 0 < t \\ 0 & \text{if } t \leq 0 . \end{cases}$$

Now we need a crucial geometric property of this homotopy H.

Lemma 6.2.1 *Let $F : \mathbf{R}^N \to \mathbf{R}^N$ be a C^2 mapping. Suppose that $F(0) = 0$ and that the Jacobian determinant of F is nonzero at each point. Then, for each $y \in \mathbf{R}^N$, $H^{-1}(y)$ consists of a nonempty union of closed arcs. Moreover: If $A \subseteq H^{-1}(y)$ is an arc, and if the hyperplane $t = c$ cuts A, then it does so transversely in exactly one point.*

Proof. We write $H(t, x) = (H_1(t, x), H_2(t, x), \ldots, H_N(t, x))$. Then DH is the $N \times (N + 1)$ matrix

$$\begin{pmatrix} \dfrac{\partial H_1}{\partial t} & \dfrac{\partial H_1}{\partial x_1} & \cdots & \dfrac{\partial H_1}{\partial x_N} \\ \vdots & \vdots & & \vdots \\ \dfrac{\partial H_N}{\partial t} & \dfrac{\partial H_N}{\partial x_1} & \cdots & \dfrac{\partial H_N}{\partial x_N} \end{pmatrix} \tag{6.18}$$

By (6.17), for $0 \leq t$, the $N \times N$ matrix obtained by omitting the first column of DH is $DF(tx)$. Since DF is nonsingular everywhere, it follows that DH has full rank at every point.

Since DH has full rank at each point, it follows from the rank theorem that, for each $y \in \mathbf{R}^N$, the set $H^{-1}(y)$ consists of a union of closed arcs. These may be compact or not and some may be topological circles. By inspection, the endpoints of these arcs must lie in $\left(\{0\} \times \mathbf{R}^N\right) \cup \left(\{1\} \times \mathbf{R}^N\right)$—the boundary of $I \times \mathbf{R}^N$. In fact, one can see that

$$H^{-1}(y) \cap \left(\{0\} \times \mathbf{R}^N\right) = \{0\} \times \left\langle [DF(0)]^{-1}, y \right\rangle \tag{6.19}$$

and
$$H^{-1}(y) \cap \left(\{1\} \times \mathbf{R}^N\right) = \{1\} \times F^{-1}(y).$$

In (6.19), $[DF(0)]^{-1}$ is the inverse of the linear map $DF(0)$.

Now let $A \subseteq H^{-1}(y)$ be any arc. To each $P \in A$ we assign a continuously differentiable unit tangent vector

$$\lambda(P)\mathbf{e}_1 + v\,, \tag{6.20}$$

where $\mathbf{e}_1$ is a unit vector pointing along the t-axis and $v = (0, a_1, \ldots, a_N)$ lies in the space orthogonal to $\mathbf{e}_1$. Since H is constant along A,

$$\langle DH(P), [\lambda(P)\mathbf{e}_1 + v]\rangle = 0.$$

Therefore
$$\lambda(P)\langle DH(P), \mathbf{e}_1\rangle = -\langle DH(P), v\rangle.$$

But (6.18) tells us that

$$\langle DH(P), v\rangle = \langle DF(tx), w\rangle,$$

where $w = (a_1, \ldots, a_N)$, $t = t(P)$, and $x = x(P)$. As a result

$$\lambda(P)\langle DH(P), \mathbf{e}_1\rangle = -\langle DF(tx), w\rangle\,. \tag{6.21}$$

It follows from (6.21) that, if the hyperplane $\{t = c\}$ cuts A at P, it does so transversely; for, if this were not the case, then it would follow that $\lambda(P) = 0$ and therefore that $|w| = 1$; as a consequence, $DF(cx)$ would be singular. That would be a contradiction. Furthermore, if the hyperplane $\{t = c\}$ were to cut A at a second point $Q \neq P$, then P and Q would divide A into three subarcs, at least one of which, call it A', would have both endpoints in the hyperplane $\{t = c\}$. Since A' is bounded (because both endpoints are in the hyperplane), there must be a maximum or minimum value of t along A' that is not equal to c. At any point on A' where that extreme value is attained, λ would equal 0, and that would be a contradiction. By the same reasoning, no A can be a topological circle. This completes the proof of the lemma. □

We next establish some further properties of the arcs which compose the preimages $H^{-1}(y)$.

Property 1: Let A be an arc contained in $H^{-1}(y)$. If an endpoint of A lies in $\{1\} \times \mathbf{R}^N$ and if a hyperplane $\{t = c\}$ cuts A, then so does the hyperplane $\{t = d\}$ for any $c \le d \le 1$. (This follows from the connectedness of A.) Similarly, if an endpoint of A lies in $\{0\} \times \mathbf{R}^N$ and if a hyperplane $\{t = c\}$ cuts A, then so does the hyperplane $\{t = d\}$ for any $0 \le d \le c$.

From Property 1, we immediately obtain the following property:

Property 2: If an endpoint of A lies in $\{1\} \times \mathbf{R}^N$ and if a hyperplane $\{t = c\}$ *does not* cut A, then neither does $\{t = d\}$ cut A for any $0 \le d \le c$. Similarly, if an endpoint of A lies in $\{0\} \times \mathbf{R}^N$ and if a hyperplane $\{t = c\}$ *does not* cut A, then neither does $\{t = d\}$ cut A for $c \le d \le 1$.

Let

$$S = \{c : \text{the hyperplane } t = c \text{ cuts } A\}.$$

Then S is a bounded subset of the real numbers. Either 0 or 1 is in S. In case 1 is in S, let $c_0 = \inf S$. If there is a real number $a > 0$ such that the part of A which lies in the closed region of $I \times \mathbf{R}^N$ between the hyperplanes $\{t = c_0\}$ and $\{t = c_0 + a\}$ is bounded then, since A is obviously closed, the hyperplane $\{t = c_0\}$ cuts A. We can reason similarly in case 0 is in S. We conclude that:

Property 3: Either the curve A is compact or else it is asymptotic to the hyperplane $\{t = c_0\}$.

We parameterize the arc A in the following way: If the hyperplane $\{t = c\}$ cuts A at P, then P is given coordinates $(c, x_1(c), \ldots, x_N(c))$. Here $x(P) = (x_1(c), \ldots, x_N(c))$. Thus we may represent λ, defined in (6.20), as a function of t:

$$\lambda(t) = \pm(1 + \dot{x}_1^2 + \cdots + \dot{x}_N^2)^{-1/2}; \tag{6.22}$$

here the sign is chosen depending on the orientation of the arc A and the accent $\cdot$ is used to denote differentiation in t.

In order to prove our first version of Hadamard's global inverse function theorem, we first must introduce some ancillary terminology from topology.

Definition 6.2.2 Let X, Y be topological spaces and $g : X \to Y$ a mapping. We say that g is *proper* if whenever $K \subseteq Y$ is compact then $g^{-1}(K) \subseteq X$ is compact.

The notion of properness bears some discussion. It is distinct from continuity. A proper mapping need not be continuous. For example, the mapping

$$f : [0, 1] \to (0, 1)$$

given by

$$f(x) = \begin{cases} x & \text{if } 0 < x < 1 \\ 1/2 & \text{if } x = 0 \text{ or } 1 \end{cases}$$

is proper but not continuous. Conversely, a continuous mapping need not be proper (exercise).

When the topological spaces in question are open domains in $\mathbf{R}^N$, then there is a useful alternative formulation of properness. Namely, let U, V be connected open sets (domains) in $\mathbf{R}^N$. Let $g : U \to V$ be a mapping. The mapping g is proper (according to Definition 6.2.2) if and only if whenever $\{x_j\} \subseteq U$ satisfies $x_j \to \partial U$ then $f(x_j) \to \partial V$. We leave the details of the equivalence of the two definitions as an exercise for the reader, or refer to Kelley [Ke 55].

Theorem 6.2.3 (Hadamard) *Let $F : \mathbf{R}^N \to \mathbf{R}^N$ be a C^2 mapping. Suppose that $F(0) = 0$ and that the Jacobian determinant of F is nonzero at each point. Further suppose furthermore that F is proper. Then F is one-to-one and onto.*

Proof. Define H as above, and consider $y \in \mathbf{R}^N$.

Let A be an arc in $H^{-1}(y)$. If A were not compact, then A would be asymptotic to a hyperplane $\{t = c_0\}$, $0 < c_0 < 1$. So there would exist sequences $t_j \to c_0$ and $x_j \in \mathbf{R}^N$ with $|x_j| \to \infty$ such that $F(t_j x_j)/t_j = y$ for all j. We conclude that $F^{-1}\left[\overline{\mathbf{B}}(c_0 y, 1)\right]$ is unbounded, contradicting the assumption that F is proper.

Because $DF(0)$ is nonsingular, we know that $H^{-1}(y)$ must contain an arc with endpoint $\left(0, \langle [DF(0)]^{-1}, y\rangle\right)$. Since that arc must be compact (as we just showed) and since by Lemma 6.2.1 each hyperplane $\{t = c\}$ must cut the arc in at most one point, the other endpoint of the arc must be in $\left(1, F^{-1}(y)\right)$. Thus F maps onto $\mathbf{R}^N$.

To see that F is one-to-one, suppose there is another distinct point in $\left(1, F^{-1}(y)\right)$ That point must be the endpoint of another distinct arc in $H^{-1}(y)$. Reasoning as above, we see that the second endpoint of this new arc must be a second distinct point in $\left(0, \langle [DF(0)]^{-1}, y\rangle\right)$, contradicting the fact that $DF(0)$ is nonsingular. □

Next we present a refined version of Hadamard's theorem.

Theorem 6.2.4 *Let $F : \mathbf{R}^N \to \mathbf{R}^N$ be a C^2 mapping. Suppose that $F(0) = 0$ and that the Jacobian determinant of F is nonzero at each point. Finally, assume that there is a constant K such that*

$$|\langle [DF(x)]^{-1}, v\rangle| \leq K|v| \tag{6.23}$$

holds for all $x \in \mathbf{R}^N$, and vectors $v \in \mathbf{R}^N$. Then F is a C^2 diffeomorphism.

Proof. Proceeding as in the proof of Theorem 6.2.3, define H as before, consider $y \in \mathbf{R}^N$, and let A be an arc in $H^{-1}(y)$. We need to show that A is compact.

If A were not compact, then A would be asymptotic to a hyperplane $\{t = c_0\}$, $0 < c_0 < 1$. Let A be parametrized by $(t, x(t))$ either for $0 \leq t < c_0$ or for $c_0 < t \leq 1$. We have

$$F(tx(t)) = ty$$

so, after differentiating with respect to t, we obtain

$$\langle DF(tx(t)), t\dot{x}(t)\rangle + \langle DF(tx(t)), x(t)\rangle = y. \tag{6.24}$$

Since DF is everywhere invertible, we can rewrite (6.24) as

$$\dot{x} + x = \langle [DF(tx)]^{-1}, y\rangle. \tag{6.25}$$

Since $|x(t)| \to \infty$ as $t \to c_0$ and since, according to (6.23),

$$|\langle [DF(tx)]^{-1}, y\rangle|$$

is uniformly bounded by $K\,|y|$, there is an $\epsilon > 0$ so that $0 < |t - c_0| < \epsilon$ implies $|\langle [DF(tx)]^{-1}, y\rangle| \leq |x(t)|/2$. Thus we have

$$\dot{x} \cdot x + x \cdot x = \langle [DF(tx)]^{-1}, y\rangle \cdot x \leq \tfrac{1}{2} x \cdot x. \tag{6.26}$$

Set $m(t) = |x(t)|$. From (6.26) it follows that

$$\dot{m}(t) \leq -\tfrac{1}{2} m(t) \tag{6.27}$$

holds for $0 < |t - c_0| < \epsilon$. Choosing a value $t_0 \neq c_0$ at distance less than ϵ from c_0, we can integrate (6.27) to conclude that

$$0 \leq m(t) \leq m(t_0) e^{-\frac{1}{2}|t - t_0|}$$

holds for t between t_0 and c_0, contradicting the assumption that $|x(t)| \to \infty$ as $t \to c_0$.

The remainder of the proof of the theorem follows that of Theorem 6.2.3. □

As an application of Hadamard's theorem, we will give a proof of the fundamental theorem of algebra.

Theorem 6.2.5 (Fundamental Theorem of Algebra) *If $p(z)$ is a nonconstant complex polynomial, then there is a $z_0 \in \mathbb{C}$ for which $p(z_0) = 0$.*

Proof. Assume that $p(z)$ is a polynomial of degree $n \geq 1$. Let P be an anti-derivative for p (choose and fix one such anti-derivative). Of course P will be a polynomial of degree $n + 1$. Consider a large circle centered at the origin. The image under P of such a circle will be a curve that encircles the origin $n + 1 \geq 2$ times. Such a curve must cross itself, for if it did not, then it would be a simple closed curve, so by the Jordan curve theorem it would enclose a cell. Thus it could not encircle the origin the two or more times required. In summary, the mapping $z \mapsto P(z)$ is not one-to-one.

We conclude that there must be a $z_0 \in \mathbb{C}$ such that $P'(z_0) = p(z_0) = 0$. This is so because, if $P'(z) = p(z)$ never vanished, then Hadamard's theorem (the properness follows because $P(z)$ behaves like its leading term when z is large) would imply that the mapping P is a diffeomorphism; and we know it is not. □

Remark 6.2.6 The reader who feels the preceding proof is too heuristic may appeal to the following two facts from elementary complex analysis:

> *For a polynomial of degree d, the image of any sufficiently large circle centered at the origin has winding number $\pm d$ about 0.*

> *The winding number of a Jordan curve is ± 1 or 0 about any point not on the curve.*

Another form of Hadamard's global inverse function theorem, which we discuss next, allows more freedom in the choice of domain and range of the function, but involves more sophisticated topological considerations. Because more topological background is involved, we will not prove the result here. A proof can be found in Gordon [Go 72].

We recall what it means for a topological space to be simply connected.

Definition 6.2.7 Let X be a topological space. We say the X is *simply connected* if every closed curve in X can be contracted to a point. More precisely, the requirement is that whenever $\phi : [0, 1] \to X$ is a closed curve (that is, ϕ is continuous and $\phi(0) = \phi(1)$) there exists a continuous function $H : [0, 1] \times [0, 1] \to X$ such that

(1) $H(t, 0) = \phi(t)$, for all $t \in [0, 1]$,

(2) $H(0, u) = H(1, u) = p$, for all $u \in [0, 1]$, where $p = \phi(0) = \phi(1)$, and

(3) $H(t, 1) = p$, for all $t \in [0, 1]$.

Euclidean space $\mathbf{R}^N$ is simply connected for every choice of $N \geq 1$, but the sphere $\mathbb{S}^N$ is simply connected only when N is greater than or equal to 2. An annulus or torus is never simply connected. The fact that is relevant to our application is that for $N \geq 3$ a point can be deleted from $\mathbf{R}^N$ and the space remains simply connected. In contrast, $\mathbf{R}^2 \setminus \{p\}$ is not simply connected.

Theorem 6.2.8 (Hadamard) *Let M_1 and M_2 be smooth, connected N-dimensional manifolds and let $f : M_1 \to M_2$ be a C^1 function. If*

(1) *f is proper,*

(2) *the Jacobian of f vanishes nowhere, and*

(3) *M_2 is simply connected,*

then f is a homeomorphism.

Based on ideas in Gordon [Go 77], we now give an application of this theorem of Hadamard to a fundamental question of differential topology.

The real line, $\mathbf{R}^1$, is canonically identified with the field of real numbers, and $\mathbf{R}^2$ is identified with the field of complex numbers via the Argand diagram (1806). Hamilton (1805–1865) showed that $\mathbf{R}^4$ can be endowed with the algebraic structure of the quaternions (1843), and Cayley (1821–1895) showed that $\mathbf{R}^8$ carries the algebraic structure known as the octonions or Cayley numbers (1845). It is natural to ask which other Euclidean spaces can be equipped with a field structure, or a division ring structure, or something similar.[1]

[1] Of course the algebraic operations must be required to be smooth or every R^n can be made a field trivially by using the one-to-one, onto mapping from set theory that demonstrates that R^1 and R^n have the same cardinality.

While these questions were investigated in the nineteenth century (see Kline [Kl 72]), we can now say definitively—and this is one of the great triumphs of twentieth-century mathematics—that the examples above form the complete list: Only dimensions 1, 2, 4, 8 can have the sort of structure we seek. And the standard structures—reals, complex numbers, quaternions, and Cayley numbers—are the unique such structures. The precise result is that $\mathbf{R}^n$ is a (possibly nonassociative) normed division algebra only for $n = 1, 2, 4, 8$. This result is the work of J. Frank Adams [Ad 60], and it relies on Adams's analysis of the Steenrod algebra. While it would not be appropriate to discuss Steenrod algebras here, we can in fact use the ideas developed in the present section to show that there is no (commutative) division ring structure on $\mathbf{R}^3$. We now turn to that task.

In the following theorem, we use the arbitrarily chosen symbol $\diamond$ to denote a hypothetical operation on $\mathbf{R}^N$.

Theorem 6.2.9 *For $N \geq 3$ there is no operation $\diamond$ of multiplication on $\mathbf{R}^N$ which satisfies the following axioms (for $x, y, z \in \mathbf{R}^N$ and $\lambda \in \mathbf{R}$)*

(1) $x \diamond (\lambda y) = (\lambda x) \diamond y = \lambda (x \diamond y)$,

(2) $x \diamond (y + z) = x \diamond y + x \diamond z$,

(3) $x \diamond y = 0 \implies x = 0$ *or* $y = 0$,

(4) $x \diamond y = y \diamond x$.

Proof. By Axioms (1), (2), and (4), we see that if $x = (x_1, x_2, \ldots, x_N)$ and $y = (y_1, y_2, \ldots, y_N)$, then

$$x \diamond y = \sum_{i,j=1}^{N} x_i y_j \left(\mathbf{e}_i \diamond \mathbf{e}_j\right), \tag{6.28}$$

where $\mathbf{e}_i$, $\mathbf{e}_j$ are elements of the standard basis.

Let $G : \mathbf{R}^N \to \mathbf{R}^N$ be given by $G(x) = x \diamond x$. By (6.28), G is smooth and we can calculate that

$$\langle DG(x), v \rangle = 2x \diamond v, \tag{6.29}$$

so by Axiom (3), DG is nonsingular on $\mathbf{R}^N \setminus \{0\}$.

The restriction of G to $g : \mathbf{R}^N \setminus \{0\} \to \mathbf{R}^N \setminus \{0\}$ is continuous and is easily shown to be proper. By Axiom (3) and (6.29), we conclude that the Jacobian of g vanishes nowhere on $\mathbf{R}^N \setminus \{0\}$. Thus g is a homeomorphism by Theorem 6.2.8 above. But Axiom (1) entails that $(-x) \diamond (-x) = x \diamond x$, so this is obviously absurd. This contradiction establishes the result. □

6.3 The Implicit Function Theorem via the Newton–Raphson Method

We have already presented the implicit function theorem from the classical point of view of estimates (Section 3.3) and from the point of view of fixed point theory (Section 3.4). One of the most useful approaches to the theorem is by way of the so-called Newton–Raphson (Joseph Raphson: 1648–1715) method. This approach has the advantage of universality: it works in any norm. As a result, one sees immediately that the implicit function theorem is valid not only in C^k but in other function spaces such as Sobolev spaces, Lipschitz spaces, Besov spaces, and so forth. We follow the approach of Cesari [Ce 66]. We begin with the setup for our theorem.

Notation 6.3.1

(1) Let Y be a Hausdorff, locally convex, topological vector space.

(2) Suppose Y has the property that if a sequence $\{y_n\}_{n=1,2,\ldots}$ is such that,

$$\begin{aligned}&\text{for any neighborhood } V \text{ of the origin, there is an } M\\ &\text{for which } M \le i \text{ and } M \le j \text{ imply } y_i - y_j \in V,\end{aligned} \tag{6.30}$$

then the sequence converges to some $y \in Y$.

(3) Let $Y_0 \subseteq Y$.

(4) Let F be another linear space. Suppose that we are given a functional $f : Y_0 \to F$. We assume an "approximate" solution y_0 of the equation $f(y) = 0$ is given.

(5) Let $\mathcal{V} = \{V_\alpha\}_{\alpha \in \mathcal{A}}$ be a neighborhood basis of $0 \in Y$, that is, every neighborhood of $0 \in Y$ contains an element of $\mathcal{V}$. Suppose that if $V \in \mathcal{V}$ and $\lambda > 0$, then $\lambda V \in \mathcal{V}$.

(6) We further suppose that each element of $\mathcal{V}$ is balanced, absorbent, and convex.[2]

(7) Let S be a closed, convex subset of Y_0 such that $S_0 \equiv S - y_0$ is balanced.

Definition 6.3.2 Condition (2) will hold whenever Y is a complete metric space,[3] so by analogy we will say that a sequence $\{y_n\}_{n=1,2,\ldots}$ that satisfies (6.30) is a *generalized Cauchy sequence*.

---

[2]A set B is *balanced* if $cB \subseteq B$ holds for all scalars c with $|c| \le 1$. A set A is *absorbent* if $Y = \bigcup_{t>0} tA$.

[3]There is a generalization of the notion of "completeness" that can be applied to non-metrizable spaces. Our condition (2) also will hold for such complete spaces. The interested reader should consult Arkhangel'skiĭ and Fedorchuk [AF 90], Bourbaki [Bo 89], Page [Pa 78], or Zeidler [Ze 86]. The latter reference is specifically intended for the context of functional analysis.

Theorem 6.3.3 *We use the notation and machinery from above. Assume that there are numbers k_0, k with $0 \leq k_0 < 1$, $0 \leq k \leq 1 - k_0$ and linear operators $B : F \to Y$ and $A : Y \to F$ with the following properties: We suppose that B has trivial null space and that, whenever $y_1, y_2 \in S$ and $y_1 - y_2 \in V \in \mathcal{V}$, it holds that*

$$B[f(y_1) - f(y_2) - A(y_1 - y_2)] \in k_0(S_0 \cap V), \tag{6.31}$$

$$Bf(y_0) \in kS_0, \tag{6.32}$$

$$BA = I. \tag{6.33}$$

If $T : S \to Y$ is given by

$$Ty = y - Bf(y) \tag{6.34}$$

for $y \in S$, then there is one and only one element $\psi \in S$ with $f(\psi) = 0$.

In the proof, we shall use a fixed point construction to establish the existence of ψ. After proving the theorem in detail, we shall put the result in context and illustrate it with some examples.

Proof. First note that the identity $\psi = T\psi$ implies, by (6.34), that $Bf(\psi) = 0$. Since the null space of B is zero, we conclude that $f(\psi) = 0$. Conversely, if $f(\psi) = 0$ with $\psi \in S$ then $\psi = T\psi$. In conclusion, a fixed point of T is just the same as a root of f.

Properties (6.34) and (6.32) tell us that

$$Ty_0 - y_0 = -Bf(y_0) \in kS_0. \tag{6.35}$$

If $y_1, y_2 \in S$ and $V_0 \in \mathcal{V}$, then the absorbancy property tells us that there is a $\lambda > 0$ such that $y_1 - y_2 \in \lambda V_0$. Furthermore, we have $\lambda V_0 \in \mathcal{V}$. Now, taking $V = \lambda V_0$ and noting that $BA = I$, we see that (6.31), (6.33), and (6.34) tell us that

$$Ty_1 - Ty_2 = -B[f(y_1) - f(y_2) - A(y_1 - y_2)] \in k_0 S_0. \tag{6.36}$$

Trivially,

$$Ty - y_0 = [Ty - Ty_0] + [Ty_0 - y_0] \tag{6.37}$$

holds for each $y \in S$. By (6.36) the first term on the right-hand side of (6.37) lies in $k_0 S_0$, and, by (6.35), the second term on the right-hand side of (6.37) lies in kS_0. Thus we see that $Ty - y_0$ lies in $(k + k_0)S_0 \subseteq S_0$. As a result, $T : S \to S$.

Now we set up an iteration scheme in the spirit of the Newton–Raphson method. Let $y_{n+1} = Ty_n$ for $n = 0, 1, 2, \ldots$. Of course y_0 is already given. Then $y_n \in S$ for all n. Let $V \in \mathcal{V}$. Again, by absorbancy, there is a $\lambda > 0$ such that $y_1 - y_0 \in \lambda V$. And of course $\lambda V \in \mathcal{V}$. Our key "estimate" is to show that

$$y_{n+1} - y_n \in k_0^n \lambda V, \qquad n = 0, 1, 2, \ldots. \tag{6.38}$$

This we now do.

The relationship (6.38) is plainly true when $n = 0$. Suppose inductively that (6.38) has been established for $0, 1, \ldots, n-1$. We next prove the assertion for n. In point of fact, we have from (6.28) that

$$y_{n+1} - y_n = -B[f(y_n) - f(y_{n-1}) - A(y_n - y_{n-1})] \in k_0\big(k_0^{n-1}\lambda V\big) = k_0^n \lambda V\,.$$

Thus (6.38) is now proved for every n. Iterating, we have that

$$y_{n+p} - y_n = (y_{n+1} - y_n) + (y_{n+2} - y_{n+1}) + \cdots + (y_{n+p} - y_{n+p-1})\,. \tag{6.39}$$

We see that $y_{n+p} - y_n$ lies in

$$k_0^n \lambda V + k_0^{n+1}\lambda V + \cdots + k_0^{n+p-1}\lambda V \subseteq (1-k_0)^{-1} k_0^n \lambda V\,, \tag{6.40}$$

where, of course, $k_0 < 1$. Now, given any $V \in \mathcal{V}$, there is some $\tilde{n}$ such that if $n \geq \tilde{n}$ and $p \geq 0$, then $(1-k_0)^{-1}k_0^n\lambda < 1/2$ and

$$y_{n+p} - y_n \in (1/2)V \subseteq (1/2)\overline{V} \subseteq V.$$

We conclude that $\{y_n\}$ is a generalized Cauchy sequence in Y. So by hypothesis, $\psi = \lim_{n\to\infty} y_n$ exists.

Note that $\psi \in Y$ and $\psi \in S$. Finally, (6.40) implies that $\psi - y_n$ belongs to the closure of $[(1-k_0)^{-1}k_0^n\lambda]V$ when $n \geq \tilde{n}$. Thus, for $n > \tilde{n}$, we see that

$$\begin{aligned} \psi - T\psi &= (\psi - y_{n+1}) + (y_{n+1} - Ty_n) + (Ty_n - T\psi) \\ &\in \frac{1}{2}V + k_0(1-k_0)^{-1}k_0^n\lambda V \\ &\subseteq \frac{1}{2}V + \frac{1}{2}V = V\,. \end{aligned}$$

Here V is an arbitrary element of $\mathcal{V}$. Since Y is Hausdorff, we conclude that $\psi - T\psi = 0$; in other words, ψ is a fixed element of T. This establishes existence. The uniqueness of ψ in S now follows from this standard argument:

If y, z are fixed points of T, then $y = T^n y$ and $z = T^n z$ for every n. If $V \in \mathcal{V}$ and λ is chosen so that $y - z \in \lambda V$, then (6.31) applied to y and to z and their iterates gives

$$y - z = T^n y - T^n z \in k_0^n \lambda V$$

for every n. We may choose n so large that $|k_0^n\lambda| < 1$. Thus $y - z \in V$ for every $V \in \mathcal{V}$. Since Y is Hausdorff, we conclude that $y - z = 0$. □

In fact, it can be shown that ψ depends continuously on parameters. Such information is crucial in many applications. We now formulate and prove such a result. Some preliminaries are required. We continue to use Notation 6.3.1, but extend it as follows:

Notation 6.3.4 Let $Y_0 \subseteq Y$ as before, let Z be a locally convex topological vector space and let $Z_0 \subseteq Z$ be any subset. Assume that $f : Y_0 \times Z_0 \to F$. Let $\mathcal{W} = \{W_\alpha\}_{\alpha \in \mathcal{A}}$ be a neighborhood basis of $0 \in Z$. We let Σ denote some closed subset of Z which is contained in Z_0.

We fix a closed, convex subset $S \subseteq Y$ as before, and begin with an initial approximate solution y_0. Assume that there are numbers k_0, k with $0 \leq k_0 < 1$ and $0 \leq k \leq 1 - k_0$ and linear operators $B : F \to Y$ and $A : Y \to F$, B having trivial null space, such that if $y_1, y_2 \in S$ and $y_1 - y_2 \in V \in \mathcal{V}$, then

$$B[f(y_1) - f(y_2) - A(y_1 - y_2)] \in k_0(S_0 \cap V), \tag{6.31}$$
$$Bf(y_0) \in kS_0, \tag{6.32}$$
$$BA = I. \tag{6.33}$$

We make the following standing hypothesis:

H1: We assume that (6.31), (6.32), and (6.33) hold uniformly for any choice of $z \in \Sigma \subseteq Z_0$ (we think here of z as a parameter and of y as the operative variable). Both operators $A = A_z$ and $B = B_z$ will, in general, depend on the variable z.

By the theorem we have already established, for every $z \in \Sigma$ there is a unique $\psi = \psi_z \in S$ such that $f(\psi_z, z) = 0$. This ψ_z is the fixed point of the map $T_z : S \to S$ defined by $T_z y \equiv y - B_z f(y, z)$. In other words, $\psi_z = T_z \psi_z$ with $\psi_z \in S$ and $z \in \Sigma$. Let $\tau : \Sigma \to S$ be defined by $\psi_z = \tau(z)$ for $z \in \Sigma$ and $\psi_z \in S$. We will invoke the following additional hypothesis:

H2: Given $V \in \mathcal{V}$ there is some $W \in \mathcal{W}$ such that if $z_1, z_2 \in \Sigma$ with $z_1 - z_2 \in W$ and if $y \in S$, then

$$B_{z_1} f(y, z_1) - B_{z_2} f(y, z_2) \in V. \tag{6.41}$$

We also assume the analogous hypothesis that, given $V \in \mathcal{V}$ and any compact subset C' of Z, there is some $W \in \mathcal{W}$ such that $z_1, z_2 \in \Sigma \cap C'$, $z_1 - z_0 \in W$, $y \in S$ implies (6.41).

Theorem 6.3.5 *Under the hypotheses H1 and H2, $\tau : \Sigma \to S$ is uniformly continuous on Σ.*

Proof. Let $V \in \mathcal{V}$ and $V_0 = \frac{1}{2}(1 - k_0)V$. By H2, there is an element $W \in \mathcal{W}$ such that $z_1, z_2 \in \Sigma$ with $z_1 - z_2 \in W$, $y \in S$ implies

$$B_{z_1} f(y, z_1) - B_{z_2} f(y, z_2) \in V_0.$$

We iteratively define

$$y_{i,n+1} = T_{z_i} y_{i,n},$$

for $n = 0, 1, 2, \ldots$ and $i = 1, 2$, with $y_{1,0} = y_{2,0} = y_0$. Then

$$y_{i,1} = T_{z_i} y_0 = y_0 - B_{z_i} f(y_0, z_i), \qquad i = 1, 2,$$

hence

$$y_{1,1} - y_{2,1} = -B_{z_1} f(y_0, z_1) + B_{z_2} f(y_0, z_2) \in V_0 .$$

We next show that

$$y_{1,n} - y_{2,n} = T_{z_1}^n y_0 - T_{z_2}^n y_0 \in (1 + k_0 + \cdots + k_0^{n-1}) V_0 , \quad n = 1, 2, \ldots . \tag{6.42}$$

This assertion is certainly true for $n = 1$. Let us assume that (6.42) is true for $1, 2, \ldots, n$ and then prove it for $n + 1$. In fact

$$\begin{aligned} y_{1,n+1} - y_{2,n+1} &= -B_{z_1} f(y_{1,n}, z_1) + B_{z_2} f(y_{2,n}, z_2) + y_{1,n} - y_{2,n} \\ &= -B_{z_1}[f(y_{1,n}, z_1) - f(y_{2,n}, z_1) - A_{z_1}(y_{1,n} - y_{2,n})] \\ &\quad -B_{z_1} f(y_{2,n}, z_1) + B_{z_2} f(y_{2,n}, z_2) . \end{aligned}$$

As a result,

$$y_{1,n+1} - y_{2,n+1} \in k_0(1 + k_0 + \cdots + k_0^{n-1}) V_0 + V_0 = (1 + k_0 + \cdots + k_0^n) V_0 .$$

Inductively, (6.42) is now proved for every n.

As $n \to \infty$, we find that $\psi_{z_1} - \psi_{z_2}$ belongs to the closure of $(1 - k_0)^{-1} V_0$, hence to the closure of $\frac{1}{2} V$. Thus $\psi_{z_1} - \psi_{z_2}$ certainly belongs to V. In summary, we have proved this: Given $V \in \mathcal{V}$, there is a $W \in \mathcal{W}$ such that if $z_1, z_2 \in \Sigma$ and $z_1 - z_2 \in W$ then $\psi_{z_1} - \psi_{z_2} \in V$. Our result is thus proved. □

In case the spaces are actually Banach spaces, the proof becomes more streamlined (see, for example Leach [Le 61]), and, in fact, one can prove a more refined result. We state here, without elaboration, one such theorem. We begin with some notation.

Let U, V be Banach spaces. Let $f : U \to V$ be a function. A *strong differential* of the function f at a point $x_0 \in U$ is a bounded linear transformation $\alpha : U \to V$ which approximates the change in f in the following strict sense: For every $\epsilon > 0$ there is a $\delta > 0$ such that if x' and x'' satisfy $\|x'\| < \delta$ and $\|x''\| < \delta$ then

$$\|f(x') - f(x'') - \alpha(x' - x'')\| < \epsilon \|x' - x''\| .$$

Now we have:

Theorem 6.3.6 *Let U, V, f be as above. Assume that $f(0) = 0$ and that f has a strong differential α at 0. Let $\beta : V \to U$ be a bounded linear transformation such that $\beta\alpha\beta = \beta$. Then there is a function $g : V \to U$ such that $g(0) = 0$, g has strong differential β at 0, and g satisfies (for y near 0) the identities:*

(1) $\beta(f(g(y))) = \beta(y)$;

(2) $\beta(\alpha(g(y))) = g(y)$;

(3) $g(f(\beta(y))) = \beta(y)$;

Any two functions g satisfying these three conditions are identical for y near 0.

It is noteworthy that this new theorem does *not* mandate continuous differentiability as in the classical implicit function theorem. In fact, we only require differentiability at a single point! Observe also that the invertibility condition on the derivative at this single point is rather weaker than the classical condition. We shall say no more about Leach's result at this time.

A technical result of the type we have been considering is best understood by way of an example, and it turns out that an entire genre of examples is now easy to describe.

Example 6.3.7 Let U, V each be the same standard function space: the Lipschitz-α functions or the $C^{k,\alpha}$ functions or the Sobolev-s functions or a Besov space or L^p space. Let $f : U \to V$ be a mapping such that $f(0) = 0$ and suppose that f has a strong differential α at 0. Further suppose that α is invertible. Then the theorem tells us that f has a local inverse near 0. Note in particular that if $f : \mathbf{R}^N \to \mathbf{R}^N$ is $C^{k,\alpha}$, in the classical sense with $k \geq 1$, at the origin and if the ordinary Jacobian of f at 0 is invertible (also in the classical sense) then the mapping $T : C^{k,\alpha} \ni \varphi \mapsto \varphi \circ f$ is a bounded mapping of $C^{k,\alpha}$ to $C^{k,\alpha}$ and has a strong differential at 0 which is invertible in the sense of our Theorem 6.3.6. As a result, the mapping T is locally invertible, which means that f is invertible in the category of $C^{k,\alpha}$. A similar analysis may be applied in Sobolev spaces and the other spaces we have mentioned. □

The example shows that Theorem 6.3.6 gives us an inverse function theorem (and hence, by our standard syllogism) an implicit function theorem, in any of our familiar Banach spaces of functions. We have seen in Sections 2.4 and 6.1 that there are also implicit function theorems in the real analytic and holomorphic categories. Such theorems can be quite useful in practice, and are not so easy to pry out of the literature.

6.4 The Nash–Moser Implicit Function Theorem

6.4.1 *Introductory Remarks*

A big open problem during the last quarter of the 19th century and the first half of the 20th century was the question of whether an arbitrary Riemannian manifold can be isometrically imbedded in Euclidean space.[4] Of course the celebrated theorem of Hassler Whitney [Wh 35] tells us that the manifold can be imbedded as a smooth "surface"; it does *not* tell us that the metric will be preserved under the imbedding. It is the metric-preserving issue that Nash's theorem addresses.

The crucial technical tool that the proof of the Nash theorem showcases is a new type of implicit function theorem. Recall that our functional analysis proof of the classical implicit function theorem (Theorem 3.4.10) used the contraction mapping fixed point theorem; and that fixed point theorem is proved by iteration. Just so, the Nash theorem is proved by an iteration procedure. In point of fact, the

---

[4]The introduction of Nash [Na 56] discusses the history of the imbedding problem for Riemannian manifolds.

proof is modeled on Newton's method from calculus. But what is special about Nash's argument is that there is a very clever "smoothing" that takes place at each stage of the iteration.

Nash's paper [Na 56] is a *tour de force* of mathematical analysis. In the introduction, Nash observed that the perturbation procedure used in the paper did not seem limited to just the imbedding problem, and that view proved correct when Jürgen Moser (in Moser [Mo 61]) isolated an implicit function theorem from Nash's celebrated paper and presented it as a succinct tool of wide applicability. As a result, the theorem is known today as the Nash–Moser implicit function theorem. It is part of the standard toolkit of geometric analysis and nonlinear partial differential equations.

In this book we shall formulate and prove a version of the Nash–Moser theorem. It would take us too far afield to attempt to discuss the isometric imbedding of Riemannian manifolds. We refer the interested reader to Nash [Na 56] and Schwartz [Sc 69]. The reader may also be interested in the extensive survey of Hamilton [Ha 82], the notes of Hörmander [Ho 77], and the expositon by Saint Raymond [SR 89]. Our presentation is based on that in Schwartz [Sc 69].

6.4.2 *Enunciation of the Nash–Moser Theorem*

Notation 6.4.1 For ℓ a nonnegative integer, we define $C^\ell(\mathbf{R}^N)$ to be the space of ℓ-times continuously differentiable functions on $\mathbf{R}^N$. We equip $C^\ell(\mathbf{R}^N)$ with the norms

$$\|u\|_r = \|u\|_{C^r} = \max_{|\alpha|\le r} \sup_{x\in\mathbf{R}^N} \|D^\alpha u(x)\|,$$

where $r \le \ell$. Clearly, if $r < \rho$, then $\|u\|_r \le \|u\|_\rho$ holds.

We can now state the theorem.

Theorem 6.4.2 *Set* $P = 61$. *Let* $\mathcal{B}^k$ *be the unit ball in the space* $C^k(\mathbf{R}^N)$ *of* k*-times continuously differentiable functions on* $\mathbf{R}^N$. *Suppose that* $T : \mathcal{B}^k \to C^{k-m}(\mathbf{R}^N)$ *for some* $0 \le m \le k$. *We make the following technical hypotheses:*

(1) *The function* T *has two continuous Fréchet derivatives (see 3.4.7), both bounded by a constant* $M \ge 1$;

(2) *There is a map* L *with domain* $\mathcal{B}^k$ *and range the space*

$$\mathcal{L}(C^k(\mathbf{R}^N), C^{k-m}(\mathbf{R}^N))$$

of bounded linear operators[5] *from* $C^k(\mathbf{R}^N)$ *to* $C^{k-m}(\mathbf{R}^N)$ *such that*

[5] In the arguments that follow, we will apply this defining property of L for various values of k. Thus we are thinking of L as taking values in a space of pseudodifferential operators, which map C^k to C^{k-m} for *every* k.

(2a) $\|L(u)h\|_{k-m} \le M\|h\|_k \qquad \forall u \in \mathcal{B}^k, h \in C^k(\mathbf{R}^N)$;

(2b) $DT(u)L(u)h = h \qquad \forall u \in \mathcal{B}^k, h \in C^{k+m}$;

(2c) $\|L(u)T(u)\|_{k+9m} \le M(1 + \|u\|_{k+10m}) \qquad \forall u \in C^{k+10m}$.

Conclusion. *If*

$$\|T(0)\|_{k+9m} \le 2^{-P} M^{-5P-2}, \tag{6.43}$$

then $T(\mathcal{B}^k)$ contains the origin.

This is a new type of theorem, different from our earlier implicit function theorems, and its statement bears some discussion.

First, we hypothesize bounds on both the first and the second derivative of the function T being studied.

Second, the function L plays the role of the inverse of the derivative of T, although this property is not formulated explicitly.

Third, the hypothesis (2b) of Theorem 6.4.2 is the familiar one about the invertibility of the "derivative" L.

Fourth, hypotheses (2a) and (2c) of Theorem 6.4.2 are new. These are the smoothness hypotheses that give us leverage in the iteration scheme.

And now let us look at the conclusion. It says that the image of the operator contains the origin. How does this relate to our more familiar notion of solving for a variable, or of inverting a mapping? Of course the answer is that there is nothing special about the origin. A simple topology/logic argument shows that the image contains a neighborhood of the origin, and this statement in turn tells us that each element near the origin has a preimage under T. And of course *that* assertion amounts to solving for one variable in terms of another.

6.4.3 First Step of the Proof of Nash–Moser

The chief technical device that will be needed in the proof of the Nash–Moser theorem is the type of iteration illustrated in the following proposition.

Proposition 6.4.3 *Let X be a Banach space, and suppose that T is a mapping whose domain is the unit ball $\mathcal{B}$ in X. Let us assume that*

(A) *The operator T has two continuous Fréchet derivatives in $\mathcal{B}$, both bounded by a constant M. For technical reasons, we assume that $M > 2$.*

(B) *There is a map L with domain $\mathcal{B}$ and range the space $\mathcal{L}(X, X)$ of bounded linear mappings of X to itself and having the properties*

(B1) $\|L(u)h\| \le M\|h\| \qquad \forall h \in X, u \in \mathcal{B}$;

(B2) $DT(u)L(u)h = h \qquad \forall h \in X, u \in \mathcal{B}$.

Conclusion. *If $\|T(0)\| < M^{-5}$, then it follows that $T(\mathcal{B})$ contains the origin.*

Remark 6.4.4 Just as the contraction mapping fixed point theorem was a paradigm for the classical implicit function theorem, so this proposition is a paradigm for the Nash–Moser theorem that we will prove a bit later.

Proof. Set

$$\lambda = \frac{3}{2}, \qquad \beta = (8/3)\log M\,. \tag{6.44}$$

It is in this argument that we will utilize the first iteration scheme. Set $u_0 = 0$. Put

$$u_{n+1} = u_n - L(u_n)T(u_n)\,. \tag{6.45}$$

We will prove inductively that

$$u_{n-1} \in \mathcal{B} \qquad P_1[n]$$

and

$$\|u_n - u_{n-1}\| \le e^{-\beta\lambda^n} \qquad P_2[n]$$

hold for all $n \ge 1$.

Condition $P_1[1]$ is trivial. Condition $P_2[1]$ simply says that

$$\|L(0)T(0)\| \le e^{-\beta\lambda} \tag{6.46}$$

and this in turn will be implied by $M\|T(0)\| \le e^{-\beta\lambda}$. We know that $\|T(0)\| \le M^{-5}$, so it suffices to show $M^{-4} \le e^{-\beta\lambda}$ or, equivalently, $\beta\lambda \le 4\log M$, and the latter follows from (6.44).

Arguing by induction, we suppose that $P_1[j]$ and $P_2[j]$ are true for all $j \le n$. Then using $P_2[j]$ (and the inequality $(3/2)^j > j/2$ that we get from the binomial theorem), we estimate

$$\begin{aligned}\|u_n\| &\le \sum_{j=1}^{n} \|u_j - u_{j-1}\| \le \sum_{j=1}^{\infty} e^{-\beta\lambda^j} \\ &\le \sum_{j=1}^{\infty} e^{-\beta(\lambda-1)j} \le \frac{e^{-\beta(\lambda-1)}}{1 - e^{-\beta(\lambda-1)}} < 1\,,\end{aligned}$$

where the last inequality holds because (6.44) implies $2\log 2 < \beta$. It follows that $P_1[n+1]$ holds. As a result, the specification (6.45) makes sense.

Now if g is any twice continuously Fréchet differentiable function then the mean-value theorem with Lagrange's remainder term may be applied to $g(u+th)$ to yield (with d^2g denoting the second Fréchet derivative of g)

$$g(u+h) = g(u) + dg(u)h + \int_0^1 (1-t)\, d^2g(u+th, h, h)\, dt\,. \tag{6.47}$$

We apply the induction hypothesis, together with the estimates in (B), to obtain

$$\begin{aligned}
\|u_{n+1} - u_n\| &= \|L(u_n)T(u_n)\| \\
&\le M\|T(u_n)\| \\
&\le M\|T(u_{n-1}) - DT(u_{n-1})L(u_{n-1})T(u_{n-1})\| \\
&\quad + M^2\|u_n - u_{n-1}\|^2 \\
&= M^2\|u_n - u_{n-1}\|^2 \\
&\le M^2 e^{-2\beta\lambda^n} .
\end{aligned}$$

Notice that, in the last *equality*, we used (B2) with $h = T(u_{n-1})$.

By (6.44) we have $M^2 = e^{3\beta/4}$, so

$$M^2 e^{-2\beta\lambda^n} = e^{\beta((3/4)-2\lambda^n)} \le e^{\beta((1/2)\lambda^n - 2\lambda^n)} = e^{-\beta\lambda^{n+1}},$$

which proves $P_2[n+1]$.

It is easy to see from $P_2[n]$, $n = 1, 2, \ldots$, that $\{u_n\}$ is a Cauchy sequence, so u_n converges to some element $u \in \mathcal{B}$ as $n \to \infty$. By (B2) and (6.45), we have

$$T(u_n) = DT(u_n)(u_n - u_{n+1})$$

hence, by $P_2[n+1]$,

$$\|T(u_n)\| \le M\|u_{n+1} - u_n\| \le Me^{-\beta\lambda^{n+1}}$$

Therefore, letting $n \to 0$, we see that $T(u) = 0$. □

6.4.4 *The Crux of the Matter*

We begin with some technical definitions and terminology.

For a constant M sufficiently large, we define a family of *smoothing operators* $S(t)$ acting on C^{k-m} functions and producing C^{k+10m} functions; these will have the following properties:

$$\begin{aligned}
&\text{(A)} & \|S(t)u\|_\rho &\le Mt^{\rho-r}\|u\|_r, & u \in C^r \\
&\text{(B)} & \|(I - S(t))u\|_r &\le Mt^{r-\rho}\|u\|_\rho, & u \in C^\rho
\end{aligned}$$

where

$$k - m \le r \le \rho \le k + 10m .$$

Note that taking $r = k-m$ and $\rho = k+10m$ in (A) justifies calling S a smoothing operator. We shall later construct a family of smoothing operators using standard tricks from mathematical analysis. For now we simply assume that such a family exists.

Proof of the Nash–Moser Theorem 6.4.2. Set

$$\lambda = \frac{3}{2}, \qquad \mu = \frac{9}{4}, \qquad \beta = \frac{8}{m}\log(2M^5) . \tag{6.48}$$

Again, we use an iteration scheme. Let $u_0 = 0$. Define

$$S_n = S(e^{\beta\lambda^n}), \tag{6.49}$$

$$u_{n+1} = u_n - S_n L(u_n) T(u_n). \tag{6.50}$$

We will prove by induction that

$$u_n \in \mathcal{B}^k, \qquad Q_1[n]$$

$$\|u_n - u_{n-1}\|_k \le e^{-\mu m \beta \lambda^n}, \qquad Q_2[n]$$

and

$$1 + \|u_n\|_{k+10m} \le e^{\mu m \beta \lambda^n} \qquad Q_3[n]$$

hold for $n = 1, 2, \ldots$

Condition $Q_1[1]$ is trivial. Notice that, by (6.50), (6.49), (A) with $\rho = k$ and $r = k - m$, (2a) of Theorem 6.4.2, and (6.43), it holds that

$$\begin{aligned}
\|u_1 - u_0\|_k &= \|S_0 L(0) T(0)\|_k \\
&= \|S(e^{\beta}) L(0) T(0)\|_k \\
&\le M\, e^{m\beta}\, \|L(0)T(0)\|_{k-m} \\
&\le M^2\, e^{m\beta}\, \|T(0)\|_k \\
&\le M^2\, e^{m\beta}\, \|T(0)\|_{k+9m} \\
&\le 2^{-P} M^{-5P} e^{m\beta} = \exp\left(m\beta - P\log(2M^5)\right) < 1,
\end{aligned}$$

which is just $Q_2[1]$. Similarly, it holds that

$$\begin{aligned}
1 + \|u_1\|_{k+10m} &= 1 + \|S_0 L(0) T(0)\|_{k+10m} \\
&\le M\, e^{11m\beta}\, \|L(0)T(0)\|_{k-m} \\
&\le M^2\, e^{11m\beta}\, \|T(0)\|_k \\
&\le M^2\, e^{11m\beta}\, \|T(0)\|_{k+9m} \\
&\le 2^{-P} M^{-5P} e^{11m\beta}.
\end{aligned}$$

As a result, we have

$$\begin{aligned}
(1 + \|u_1\|_{k+10m}) e^{-\mu m \beta \lambda} &\le 2^{-P} M^{-5P} e^{(11-\mu\lambda)m\beta} \\
&= \exp\left(\tfrac{P}{8} m\beta - P\log(2M^5)\right) \le 1,
\end{aligned}$$

which gives us $Q_3[1]$.

Now, assume that $Q_2[j]$ is true for $j \le n$. Then using $\lambda^j \ge (\lambda - 1)j$, $j = 1, 2, \ldots$, we estimate

$$\begin{aligned}\|u_n\|_k &\leq \sum_{j=1}^{n} \|u_j - u_{j-1}\|_k \\ &\leq \sum_{j=1}^{\infty} e^{-\mu m\beta\lambda^j} \\ &\leq \sum_{j=1}^{\infty} e^{-\mu m\beta(\lambda-1)j} = \frac{e^{-\mu m\beta(\lambda-1)}}{1-e^{-\mu m\beta(\lambda-1)}} \leq 1,\end{aligned}$$

where the last inequality holds because $\log 2 \leq \mu m\beta(\lambda - 1)$. Thus, $Q_1[n]$ holds.

Next, suppose that $Q_1[j]$, $Q_2[j]$, $Q_3[j]$ are true for $j \leq n$. Then using (6.50), (A), (2a) of Theorem 6.4.2, (6.47) as in the proof of Proposition 6.4.3, $Q_2[n]$, the bound on DT, (B), and (2c) of Theorem 6.4.2, we estimate

$$\begin{aligned}
&\|u_{n+1} - u_n\|_k \\
&= \|S_n L(u_n) T(u_n)\|_k \\
&\leq Me^{m\beta\lambda^n} \|L(u_n)T(u_n)\|_{k-m} \\
&\leq M^2 e^{m\beta\lambda^n} \|T(u_n)\|_k \\
&\leq M^2 e^{m\beta\lambda^n} \|T(u_{n-1}) - DT(u_{n-1})S_{n-1}L(u_{n-1})T(u_{n-1})\|_k \\
&\quad + M^3 e^{m\beta\lambda^n} \|u_n - u_{n-1}\|_k^2 \\
&\leq M^2 e^{m\beta\lambda^n} \|DT(u_{n-1})(I - S_{n-1})L(u_{n-1})T(u_{n-1})\|_k \\
&\quad + M^3 e^{m\beta\lambda^n} e^{-2\mu m\beta\lambda^n} \\
&\leq M^3 e^{m\beta\lambda^n} \|(I - S_{n-1})L(u_{n-1})T(u_{n-1})\|_k \\
&\quad + M^3 e^{m\beta\lambda^n} e^{-2\mu m\beta\lambda^n} \\
&\leq M^3 e^{m\beta\lambda^n} \left[Me^{-9m\beta\lambda^{n-1}} \|L(u_{n-1})T(u_{n-1})\|_{k+9m} + e^{-2\mu m\lambda^n} \right] \\
&\leq M^3 e^{m\beta\lambda^n} \left[e^{-9m\beta\lambda^{n-1}} M^2 (1 + \|u_{n-1}\|_{k+10m}) + e^{-2\mu m\beta\lambda^n} \right] \\
&\leq M^3 e^{m\beta\lambda^n} \left[M^2 e^{-9m\beta\lambda^{n-1}} e^{\mu m\beta\lambda^{n-1}} + e^{-2\mu m\beta\lambda^n} \right] \\
&\leq M^3 \left[M^2 e^{m\beta\lambda^{n-1}(\mu - 9 + \lambda)} + e^{m\beta\lambda^n(1-2\mu)} \right] \\
&= M^3 (M^2 + 1) e^{-(21/4)m\beta\lambda^{n-1}} \\
&\leq 2M^5 e^{-(21/4)m\beta\lambda^{n-1}} \leq e^{-\mu m\beta\lambda^{n+1}}
\end{aligned}$$

where the last inequality is true because

$$\log(2M^5) \le \frac{3}{16} m\beta \le \lambda^{n-1} m\beta[(21/4) - \mu\lambda^2].$$

To prove $Q_3[n+1]$, we observe by (6.50), (A), (2c) of Theorem 6.4.2, and $Q_3[j]$, $j = 1, 2, \ldots, n$, that

$$\begin{aligned}
1 + \|u_{n+1}\|_{k+10m} &\le 1 + \sum_{j=0}^{n} \|S_j L(u_j) T(u_j)\|_{k+10m} \\
&\le 1 + M \sum_{j=0}^{n} e^{m\beta\lambda^j} \|L(u_j) T(u_j)\|_{k+9m} \\
&\le 1 + M^2 \sum_{j=0}^{n} e^{m\beta\lambda^j} (1 + \|u_j\|_{k+10m}) \\
&\le 1 + M^2 \sum_{j=0}^{n} e^{m\beta(1+\mu)\lambda^j}
\end{aligned}$$

As a result, we have

$$\begin{aligned}
&(1 + \|u_{n+1}\|_{k+10m}) e^{-\mu m\beta\lambda^{n+1}} \\
&\le e^{-\mu m\beta\lambda^{n+1}} + M^2 e^{m\beta\lambda^n(1+\mu-\mu\lambda)} + M^2 \sum_{j=0}^{n-1} e^{m\beta[(1+\mu)\lambda^j - \mu\lambda^{n+1}]} \\
&\le e^{-\mu m\beta\lambda^2} + M^2 e^{-(1/8)m\beta\lambda^n} + M^2 \sum_{j=0}^{\infty} e^{m\beta\lambda^j\{1+\mu-\mu\lambda^2\}} \\
&= e^{-\mu m\beta\lambda^2} + M^2 e^{-(1/8)m\beta\lambda} + M^2 \sum_{j=0}^{\infty} e^{-(29/16)m\beta\lambda^j} \\
&\le e^{-(81/16)m\beta} + M^2 e^{-(3/16)m\beta} + M^2 \sum_{j=0}^{\infty} e^{-(29/16)m\beta(1-\lambda)j} \\
&= (2M^5)^{-81/2} + M^2 (2M^5)^{-3/2} + M^2 \frac{e^{-(29/16)m\beta(1-\lambda)}}{1 - e^{-(29/16)m\beta(1-\lambda)}} \\
&= (2M^5)^{-81/2} + M^2 (2M^5)^{-3/2} + M^2 \frac{(2M^5)^{-29/4}}{1 - (2M^5)^{-29/4}} < 1.
\end{aligned}$$

Thus $Q_3[n+1]$ follows, and our induction is complete.

Finally, the proof concludes as in that of Proposition 6.4.3. □

6.4.5 *Construction of the Smoothing Operators*

We shall construct the requisite smoothing operators on Euclidean space. Let $\widehat{a}$ be a C^∞ function with compact support in $\mathbf{R}^N$, supported in $B(0, 1)$ and identically equal to 1 near the origin. Let a be the inverse Fourier transform of $\widehat{a}$. That is,

$$a(x) = c_N \int_{\mathbf{R}^N} e^{ix\cdot\xi}\,\widehat{a}(\xi)\,d\xi\,.$$

Then of course $|a(x)| \le C$ and, integrating by parts, we see that

$$|a(x)| = \left|\frac{1}{x^\gamma}\int_{\mathbf{R}^N} e^{ix\cdot\xi}\,D^\gamma\widehat{a}(\xi)\,d\xi\right|$$

for any multiindex γ. As a result,

$$|a(x)| \le C_\gamma(1+|x|)^{-|\gamma|}\,.$$

The same calculation may be performed on any derivative of a. We find that

$$|D^\alpha a(x)| \le C_{\alpha,K}(1+|x|)^{-K}\,.$$

It also holds that

$$\int a(x)\,dx = \widehat{a}(0) = 1$$

and

$$\int x^\gamma a(x)\,dx = (-1)^\gamma \cdot D^\gamma\widehat{a}(0) = 0$$

for all multiindices γ with $|\gamma| > 0$.

Now we define

$$[S(t)u](x) = t^N \int_{\mathbf{R}^N} a(t(x-y))u(y)\,dy\,.$$

This is a standard convolution construction, commonly used in the theory of partial differential equations and harmonic analysis (see, for example, Krantz [Kr 99]).

Let us first establish property (A) of smoothing operators when $r = 0$. We need to show that

$$\|S(t)u\|_\rho \le Mt^\rho\|u\|_0\,. \tag{6.51}$$

If $|\alpha| \le \rho$ then we calculate that

$$\begin{aligned}\|D^\alpha S(t)u\|_0 &= \left\|t^{|\alpha|}t^N \int_{\mathbf{R}^N} D^\alpha a(t(x-y))u(y)\,dy\right\| \\ &\le t^{|\alpha|+N}\int_{\mathbf{R}^N} C\cdot\|u(y)\|\,dy \\ &\le Mt^{|\alpha|}\|u\|_0 \\ &\le Mt^\rho\|u\|_0\,.\end{aligned}$$

That establishes (A) when $r = 0$. Now a particular case of (A) when $r = 0$ is the case when ρ is replaced by $\rho - r$ and also $r = 0$. Thus

$$\|S(t)v\|_{\rho-r} \le Mt^{\rho-r}\|v\|_0\,.$$

Let α be any multi-index with $|\alpha| \leq r$. Then

$$\begin{aligned} \|D^\alpha S(t)u\|_{\rho-r} &= \|S(t)D^\alpha u\|_{\rho-r} \\ &\leq Mt^{\rho-r}\|D^\alpha u\|_0 \\ &\leq Mt^{\rho-r}\|u\|_r . \end{aligned}$$

Since this is true for all such α, we have

$$\|S(t)u\|_r \leq Mt^{\rho-r}\|u\|_r .$$

We tackle (B) in the same way. The case $r = 0$ reduces to

$$\|(1 - S(t))u\|_0 \leq Mt^{-\rho}\|u\|_\rho .$$

To establish this inequality, we apply Taylor's theorem with remainder:

$$\phi(1) = \sum_{k=0}^{m-1} \frac{\phi^{(k)}(0)}{k!} + \frac{1}{(m-1)!}\int_0^1 (1-\mu)^{m-1}\phi^{(m)}(\mu)\,d\mu .$$

We let $\phi(t) = u(x + ty)$ and obtain

$$\begin{aligned} u(x+y) &= \sum_{k=0}^{\rho-1} \frac{1}{k!}\left(\sum_{|\alpha|=k} y^\alpha D^\alpha u(x)\right) \\ &\quad + \frac{1}{(\rho-1)!}\sum_{|\alpha|=\rho} y^\alpha \int_0^1 (1-\mu)^{\rho-1} D^\alpha u(x+\mu y)\,d\mu . \end{aligned}$$

Thus we have

$$\begin{aligned} u(x) - S(t)u(x) &= u(x) - t^N \int_{\mathbf{R}^N} a(t(x-y))u(y)\,dy \\ &= u(x) - t^N \int_{\mathbf{R}^N} a(ty)u(x+y)\,dy \\ &= t^N \int_{\mathbf{R}^N} a(ty)u(x)\,dy \\ &\quad - t^N \int_{\mathbf{R}^N} a(ty)u(x+y)\,dy \\ &= t^N \int_{\mathbf{R}^N} a(ty)[u(x) - u(x+y)]\,dy \\ &\overset{\text{(Taylor)}}{=} -\frac{t^N}{(\rho-1)!}\sum_{|\alpha|=\rho}\int_{\mathbf{R}^N}\int_0^1 a(ty)(1-\mu)^{\rho-1}y^\alpha \\ &\quad \times D^\alpha u(x+\mu y)\,d\mu\,dy . \end{aligned}$$

The change of variable $ty = z$ yields

$$\begin{aligned} & t^N \int_{\mathbf{R}^N}\int_0^1 a(ty)(1-\mu)^{\rho-1}y^\alpha D^\alpha u(x+\mu y)\,d\mu\,dy \\ &= t^{-|\alpha|}\int_{\mathbf{R}^N}\int_0^1 a(z)(1-\mu)^{\rho-1}z^\alpha D^\alpha u(x+\mu t^{-1}z)\,d\mu\,dz . \end{aligned}$$

But now straightforward estimates give the desired conclusion, and (B) is established.

6.4.6 A Useful Corollary

We now derive a consequence of Theorem 6.4.2 that is useful in practice (and which, incidentally, is formulated more like a classical implicit function theorem).

Notation 6.4.5 For $z \in \mathbf{R}^N$ define the *translation* $\tau_z : \mathbf{R}^N \to \mathbf{R}^N$ by setting

$$\tau_z(x) = x + z .$$

Theorem 6.4.6 *Let T be a mapping from the unit ball $\mathcal{B}^k$ of $C^k(\mathbf{R}^N)$ into $C^{k-\beta}(\mathbf{R}^N)$. Assume that $T(0) = 0$. Further suppose that*

(1) *T has infinitely many continuous Fréchet derivatives.*

(2) *T is translation invariant in the sense that if $u \in C^k$ with $\|u\|_k < 1$ then*

$$T(u \circ \tau_z) = [T(u)] \circ \tau_z$$

holds for all $z \in \mathbf{R}^N$.

(3) *There is a mapping L defined on $\mathcal{B}^k$ with values in the space of bounded linear operators from C^k to $C^{k-\beta}$ such that $L(u)$ is translation invariant in the same sense as T (in part (2)), and such that $L(u)$ has infinitely many continuous Fréchet derivatives, and finally such that*

$$\text{(3}a\text{)} \qquad \|L(u)h\|_{k-\beta} \leq M\|h\|_k \quad \forall u \in C^k, h \in C^k ,$$

$$\text{(3}b\text{)} \qquad DT(u)L(u)h = h \quad \forall u, h \in C^k .$$

Conclusion: *The set $T(\mathcal{B}^k)$ contains a neighborhood of the origin.*

Proof. Since T is translation invariant, it commutes with derivatives. Thus if we apply T to a function in C^m, $m > k$, then we obtain a function in $C^{m-\beta}$. Similarly, $L(u)$ can be considered to be a function whose domain is the unit ball of C^m and with range $C^{m-\beta}$. We also have the relations

$$\text{(3a)}^* \qquad \|L(u)h\|_{m-\beta} \leq M\|h\|_m \quad \forall u \in C^m, h \in C^m ,$$

$$DT(u)L(u)h = h \quad \forall u, h \in C^{m+\beta} .$$

Now we wish to apply the Nash–Moser Theorem 6.4.2. In order to do so, we must verify part (2c) of its statement. This follows from the translation-invariance of T and L together with part (3a) of the present theorem and the boundedness of the derivatives of T. The result of applying the Nash–Moser theorem is that if a point x is sufficiently near to 0 in $C^{k-\beta}$, then there is a point in C^k whose image is x. Therefore $T(\mathcal{B}^k)$ contains a $C^{k-\beta}$-neighborhood of the origin, and hence certainly it contains a C^∞ neighborhood as well. □

Glossary

absorbent set Let Y be a Hausdorff, locally convex, topological vector space. A set $A \subseteq Y$ is absorbent if $Y = \bigcup_{t>0} tA$.

anomaly The angle between the direction to an orbiting body and the direction to its last perihelion.

Axiom of Choice A fundamental axiom of set theory that specifies the existence of choice functions.

balanced set A set B is balanced if $cB \subseteq B$ holds for all scalars c with $|c| \leq 1$.

Banach space A normed linear space that is complete in the topology induced by the norm.

Besov space A space of functions in which smoothness is measured by certain p^{th}-power integral expressions.

Borel measurable function A function f with the property that $f^{-1}(U)$ is a Borel set when U is an open set.

Borel set A set in the σ-algebra generated by the open sets.

Cauchy estimates Certain inequalities in complex variable theory that allow estimates of the derivatives of a holomorphic function in terms of the maximum size of the holomorphic function.

Cauchy–Kowalewsky theorem A general theorem about the existence and uniqueness of real analytic solutions of partial differential equations with real analytic data.

Cauchy–Riemann equations A pair of linear partial differential equations that characterize holomorphic functions.

complex analytic function A function of a complex variable that has a convergent power series expansion (in powers of z) about each point of its domain.

continuation method A method for solving nonlinear equations by deforming the given equation to a simpler equation.

contraction A mapping F of a metric space (X, d) with the property that that there is a constant $0 < c < 1$ such that $d(F(x), F(y)) \leq c \cdot d(x, y)$ for all $x, y \in X$.

contraction mapping fixed point principle A theorem that specifies that a contraction on a complete metric space will have a fixed point.

convex homotopy A homotopy between functions F and F_0 given by the formula $H(t, x) = (1 - t)F_0(x) + tF(x)$.

decomposition theorem The result that any smooth mapping may be written as the composition of primitive mappings and linear operators which are either the identity or which exchange two coordinates.

differentiability in a Banach space See *Fréchet differentiability* or *Gâteaux differentiability*.

Dini's inductive proof of the implicit function theorem A proof of the implicit function theorem that proceeds by induction on the number of dependent variables.

distance function In a metric space, the function that specifies the distance between any two points.

division ring A ring A such that $1 \neq 0$ and such that every non-zero element is invertible.

eccentric anomaly The quantity E in Kepler's equation $E = M + e\sin(E)$, where M is the mean anomaly and e is the eccentricity.

eccentricity A numerical parameter e that specifies the shape of an ellipse.

Euler step One iteration in Euler's method.

explicit function A function that is given by a formula of the form $y = f(x)$.

Fréchet derivative A form of the derivative in a Banach space.

function Let X, Y be sets. A function from X to Y is a subset f of $X \times Y$ such that (i) for each $x \in X$ there is a $y \in Y$ such that $(x, y) \in f$ and (ii) if $(x, y) \in f$ and $(x, y') \in f$ then $y = y'$.

Fundamental Theorem of Algebra The theorem that guarantees that any polynomial of degree at least one, and having complex coefficients, will have a complex root.

Gâteaux derivative A form of the directional derivative in a Banach space.

generalized distance function A modified distance function that is smoother than the canonical Euclidean distance function.

global homotopy A homotopy between a function F and the shifted function $F - F(x_0)$ given by $H(t,x) = F(x) - (1-t)F(x_0)$.

global inverse function theorem See *Hadamard's theorem.*

Hadamard's formula A formula for the radius of convergence of a power series that is determined by the root test.

Hadamard's theorem A global form of the inverse function theorem, i.e., one that yields a global rather than a local inverse.

holomorphic function A function of a complex variable that has a complex derivative at each point of its domain. Equivalently, a function with a complex power series expansion about each point of its domain. Equivalently, a function that satisfies the Cauchy–Riemann equations.

homotopy method See *continuation method.*

homotopy A continuous deformation of curves in a topological space.

imbedding method See *continuation method.*

implicit differentiation A method for differentiating a function that is given implicitly.

implicit function theorem paradigm A general conceptual framework for implicit function theorems.

implicit function theorem A theorem that gives sufficient conditions on an equation for the solving for some of the variables in terms of the others.

implicit function A function that is given by an equation, but not explicitly.

inverse function theorem A theorem that gives sufficient conditions for the local invertibility of a mapping.

Jacobian determinant The determinant of the Jacobian matrix. If the Jacobian matrix is DG, then the Jacobian determinant is denoted by det DG.

Jacobian matrix The matrix of first partial derivatives of a mapping G, denoted by DG.

Kepler's equation An equation that relates the mean anomaly and the eccentric anomaly of a planetary orbit.

Lagrange expansion A power series expansion used to evaluate an inverse function.

Lagrange inversion theorem See *Lagrange expansion.*

Legendre transformation A change of coordinates that puts the Hamiltonian in a normalized form.

Lipschitz mapping A mapping F on a metric space (X, d) with the property that $d(F(x), F(y)) \leq C \cdot d(x, y)$.

Lipschitz space A space of Lipschitz functions.

local coordinates A method of specifying Euclidean-like coordinates in a neighborhood on a manifold.

majorant A power series whose coefficients bound above the moduli of the corresponding coefficients of another power series.

manifold A topological space that is locally homeomorphic to some Euclidean space.

maximum modulus theorem The theorem that says that a holomorphic function will never assume a local absolute maximum value in the interior of a domain.

mean anomaly See *anomaly* and *eccentric anomaly.*

method of majorants The technique of using majorants to prove convergence of a power series.

metric space A space in which there is a notion of distance satisfying certain standard axioms.

Nash–Moser implicit function theorem A sophisticated implicit function theorem proved by a complicated scheme of iteration and smoothing.

Newton diagram A graphical device for determining the qualitative behavior of a locus of points in the plane.

Newton polygon See *Newton diagram.*

Newton's method An iterative method for finding the roots of a function.

Newton–Raphson formula The iterative formula that occurs in Newton's method.

parametrization of a surface A method of assigning coordinates to a surface by means of a local mapping from Euclidean space.

Picard's iteration technique The recursive technique used in proving Picard's theorem.

Picard's theorem A fundamental theorem about the existence and uniqueness of solutions to a very general class of ordinary differential equations.

positive reach The property of a set S specifying that there is a neighborhood U of S such that each point of U has a unique nearest point in S.

primitive mapping Let $E \subseteq \mathbf{R}^N$ be open and $F : E \to \mathbf{R}^N$ a mapping. Assume that, for some fixed, positive integer j,

$$\mathbf{e}_i \cdot F(x) = \mathbf{e}_i \cdot \mathbf{x}$$

for all $\mathbf{x} \in E$ and $i \neq j$. Then F is called primitive.

proper map A map $f : X \to Y$ of topological spaces with the property that the inverse image of any compact set is compact.

Puiseux series A series of powers of a given fractional power of x.

rank theorem An implicit function type theorem that says that certain types of smooth mappings give a foliation of the domain over the range.

real analytic function A function of one or more real variables that has a local power series expansion (in powers of the x_j's) about each point of its domain.

regularly imbedded submanifold of $\mathbf{R}^N$ A manifold that can be realized, by way of a smooth mapping, as a surface in Euclidean space.

Rouché's theorem A theorem based on the argument principle of complex analysis that enables the counting of the roots of a holomorphic function.

Sard's theorem A theorem about the size of the set of critical values of a smooth mapping.

Schauder fixed point theorem A fixed point theorem for Banach spaces. Of wide utility in the theory of differential equations.

Schwarz–Pick lemma An invariant generalization of the Schwarz lemma from complex analysis.

set-valued directional derivative A generalization of the Gâteaux derivative.

signed distance function A distance function to a hypersurface that takes positive values on one side of the surface and negative values on the other side. Signed distance functions tend to be smoother than positive distance functions.

simply connected space A space in which each curve is homotopic to a point.

smooth surface A surface in space that has no singularities.

smoothing operator A linear operator that assigns to each element of a function space a smoother function.

Sobolev space A space of functions in which smoothness is measured by Lebesgue norms instead of supremum norms.

Steenrod algebra An algebraic construct in topology·involving cohomology operations.

straightening of a surface The process of applying a diffeomorphism to map a surface to a linear subspace.

strong differential A generalization of the Fréchet derivative.

true anomaly See *anomaly* and *eccentric anomaly*.

tubular neighborhood A neighborhood of a given set that is constructed as the union of normal segments.

Weierstrass polynomial The sort of polynomial that is used in the statement of the Weierstrass preparation theorem.

Weierstrass preparation theorem A theorem that says that a holomorphic function of several variables, suitably normalized, can be written as a polynomial in one of its variables.

Bibliography

John Frank Adams

[Ad 60] On the non-existence of elements of Hopf invariant one, *Annals of Mathematics* **72** (1960), 20–104.

Eugene L. Allgower and Kurt Georg

[AG 90] *Numerical Continuation Methods: An Introduction*, Springer-Verlag, Berlin, 1990.

Alexander V. Arkhangel'skiĭ and Vitaly V. Fedorchuk

[AF 90] The basic concepts and constructions of general topology, *General Topology I*, Encyclopaedia of Mathematical Sciences **17**, Springer-Verlag, New York, 1990.

Vladimir I. Arnol'd

[Ar 78] *Mathematical Ideas of Classical Mechanics*, Springer-Verlag, New York, 1978.

Bruno Belhoste

[Be 91] *Augustin-Louis Cauchy: A Biography*, translated by Frank Ragland, Springer-Verlag, New York, 1991.

Melvyn S. Berger

[Be 77] *Nonlinearity and Functional Analysis*, Academic Press, New York, 1977.

Lipman Bers

[Be 64] *Introduction to Several Complex Variables*, Courant Institute of Mathematical Sciences, New York University, 1964.

Gilbert Ames Bliss

[Bl 13] *Fundamental Existence Theorems*, American Mathematical Society Colloquium Publications, volume 3, 1913, reprinted 1934.

Nicolas Bourbaki

[Bo 89] *General Topology*, Springer-Verlag, New York, 1989.

Augustin-Louis Cauchy

[Ca 16] Résumé d'un mémoire sur la mécanique céleste et sur un nouveau calcul appelé calcul des limites, *Œuvres Complètes*, séries 2, volume 12, Gauthier-Villars, Paris, 1916, pp. 48–112.

Lamberto Cesari

[Ce 66] The implicit function theorem in functional analysis, *Duke Mathematical Journal* **33** (1966), 417–440.

Richard Courant and David Hilbert

[CH 62] *Methods of Mathematical Physics*, volume 2, Interscience Publishers, New York, 1962.

John P. D'Angelo

[DA 93] *Several Complex Variables and the Geometry of Real Hypersurfaces*, CRC Press, Boca Raton, 1993.

Ulisse Dini

[Di 07] *Lezioni di Analisi infinitesimale*, volume 1, Pisa, 1907, pp. 197–241.

Leonard Euler

[EB 88] *Introduction to Analysis of the Infinite*, translated by John D. Blanton, Springer-Verlag, New York, 1988.

Michal Fečkan

[Fe 94] An inverse function theorem for continuous mappings, *Journal of Mathematical Analysis and Applications* **185** (1994), 118–128.

Herbert Federer

[Fe 59] Curvature measures, *Transactions of the American Mathematical Society* **93** (1959), 418–491.

[Fe 69] *Geometric Measure Theory*, Springer-Verlag, New York, 1969.

Philip M. Fitzpatrick

[Fi 70] *Principles of Celestial Mechanics*, Academic Press, New York, 1970, pp. 389–393.

Wendell H. Fleming

[Fl 77] *Functions of Several Variables*, second edition, Springer-Verlag, New York, 1977.

Robert L. Foote

[Fo 84] Regularity of the distance function, *Proceedings of the American Mathematical Society* **92** (1984), 153–155.

Stephen M. Gersten

[Ge 83] A short proof of the algebraic Weierstrass preparation theorem, *Proceedings of the American Mathematical Society* **88** (1983), 751–752.

David Gilbarg and Neil S. Trudinger

[GT 77] *Elliptic Partial Differential Equations of Second Order*, Springer-Verlag, Berlin, 1977.

[GT 83] *Elliptic Partial Differential Equations of Second Order*, second edition, Springer-Verlag, Berlin, 1983.

William B. Gordon

[Go 72] On the diffeomorphisms of euclidean space, *American Mathematical Monthly* **79** (1972), 755–759.

[Go 77] An application of Hadamard's inverse function theorem to algebra, *American Mathematical Monthly* **84** (1977), 28–29.

Edouard Jean Baptiste Goursat

[Go 03] Sur la théorie des fonctions implicites, *Bulletin de la Société Mathématique de France* **31** (1903), 184–192.

Alfred Gray

[Gr 90] *Tubes*, Addison-Wesley, Redwood City, California, 1990.

Robert E. Greene and Steven G. Krantz

[GK 97] *Function Theory of One Complex Variable*, John Wiley and Sons, New York, 1997.

Jacques Hadamard

[Ha 06a] Sur les transformations planes, *Comptes Rendus des Séances de l'Académie des Sciences, Paris*, **142** (1906), 74.

[Ha 06b] Sur les transformations ponctuelles, *Bulletin de la Société Mathématique de France* **34** (1906), 71–84.

Ernst Hairer and Gerhard Wanner

[HW 96] *Analysis by Its History*, Springer-Verlag, New York, 1996.

Richard S. Hamilton

[Ha 82] The inverse function theorem of Nash and Moser, *Bulletin of the American Mathematical Society* 7(1982), 65–222.

Earle Raymond Hedrick and Wilhelmus David Allen Westfall

[HW 16] Sur l'existence des fonctions implicites, *Bulletin de la Société Mathématique de France* **46** (1916), 1–13.

Theophil Henry Hildebrandt and Lawerence Murray Graves

[HG 27] Implicit functions and their differentials in general analysis, *Transactions of the American Mathematical Society* **29** (1927), 127–153.

Einar Hille

[Hi 59] *Analytic Function Theory*, volume 1, Ginn and Company, Boston, 1959.

Ernest William Hobson

[Ho 57] *The Theory of Functions of a Real Variable and the Theory of Fourier's Series*, volume 1, Dover, New York, 1957.

Lars Hörmander

[Ho 66] *An Introduction to Complex Analysis in Several Variables*, D. Van Nostrand Company, Princeton, 1966.

[Ho 77] *Implicit Function Theorems*, unpublished lecture notes, Stanford University, 1977.

[Ho 00] Private communication.

Harold Hotelling

[Ho 39] Tubes and sphere in n-spaces, and a class of statistical problems, *American Journal of Mathematics* **61** (1939), 440–460.

Witold Hurewicz

[Hu 64] *Lectures on Ordinary Differential Equations*, M.I.T. Press, Cambridge, Massachusetts, 1964.

John L. Kelley

[Ke 55] *General Topology*, Van Nostrand, Princeton, 1955.

Morris Kline

[Kl 72] *Mathematical Thought from Ancient to Modern Times*, Oxford University Press, New York, 1972.

Steven G. Krantz

[Kr 92] *Function Theory of Several Complex Variables*, second edition, Wadsworth, 1992.

[Kr 99] *A Panorama of Harmonic Analysis*, Mathematical Association of America, Washington, D.C., 1999.

Steven G. Krantz and Harold R. Parks

[KP 81] Distance to C^k hypersurfaces, *Journal of Differential Equations* **40** (1981), 116–120.

[KP 92] *A Primer of Real Analytic Functions*, Birkhäuser Verlag, Basel, 1992.

[KP 99] *The Geometry of Domains in Space*, Birkhäuser, Boston, 1999.

Bernd Kummer

[Ku 91] An implicit-function theorem for $C^{0,1}$ equations and parametric $C^{1,1}$ optimization, *Journal of Mathematical Analysis and Applications* **158** (1991), 35–46.

Joseph Louis Lagrange

[La 69] Nouvelle méthode pour résoudre les équations littérales par le moyen des séries, (*Mémoires de l'Académie royale des Sciences et Belles-Lettres de Berlin* **24**) *Œuvres de Lagrange*, volume 3, Gauthier-Villars, Paris, 1869, p. 25.

Ernest B. Leach

[Le 61] A note on inverse function theorems, *Proceedings of the American Mathematical Society* **12** (1961), 694–697.

Solomon Lefschetz

[Le 57] *Differential Equations: Geometric Theory*, Interscience, New York, 1957.

Warren S. Loud

[Lo 61] Some singular cases of the implicit function theorem, *American Mathematical Monthly* **68** (1961), 965–977.

John D. Miller

[Mi 84] Some global inverse function theorems, *Journal of Mathematical Analysis and Applications* **100** (1984), 375–384.

Jürgen Kurt Moser

[Mo 61] A new technique for the construction of solutions of nonlinear differential equations, *Proceedings of the National Academy of Sciences of the United States of America* **47** (1961), 1824–1831.

John F. Nash

[Na 56] The imbedding problem for Riemannian manifolds, *Annals of Mathematics* **63** (1956), 20–63.

Isaac Newton

[NW 68] *Mathematical Papers of Isaac Newton*, volume 2, edited by Derek Thomas Whiteside, Cambridge University Press, 1968.

William Fogg Osgood

[Os 03] A Jordan curve of positive area, *Transactions of the American Mathematical Society* **4** (1903), 107–112.

[Os 38] *Functions of Real Variables*, G. E. Stechert and Company, New York, 1938.

Warren Page

[Pa 78] *Topological Uniform Structures*, John Wiley and Sons, New York, 1978.

Émile Picard

[Pi 93] Sur l'application des méthodes d'approximations successives à l'étude de certaines équations différentielles, *Journal de Mathématique Pures et Appliquées*, series 4, volume 9 (1893), 217–271.

Joseph Raphson

[Ra 97] *Analysis Aequationum Universalis*, T. Braddyll, London, 1697.

Walter Rudin

[Ru 64] *Principles of Mathematical Analysis*, second edition, McGraw-Hill, New York, 1964.

Xavier Saint Raymond

[SR 89] A simple Nash-Moser implicit function theorem, *L'Enseignement Mathématique (2)* **35** (1989), 217–226.

Juliusz Pawel Schauder

[Sc 30] Der Fixpunktsatz in Funktionalräumen, *Studia Mathematica* **2** (1930), 171–180.

Jacob T. Schwartz

[Sc 60] On Nash's implicit functional theorem, *Communications on Pure and Applied Mathematics* **13** (1960), 509–530.

[Sc 69] *Nonlinear Functional Analysis*, Gordon and Breach, New York, 1969.

Dirk Jan Struik

[St 69] *A Source Book in Mathematics, 1200–1800*, Harvard University Press, Cambridge, Massachusetts, 1969.

Daniel Hobson Wagner

[Wa 77] Survey of measurable selection theorems, *Society for Industrial and Applied Mathematics Journal on Control and Optimization* **15** (1977), 859–903.

Hermann Weyl

[We 39] On the volume of tubes, *American Journal of Mathematics* **61** (1939), 461–472.

Hassler Whitney

[Wh 35] Differentiable manifolds, *Annals of Mathematics* **37** (1935), 645–680.

Edmund Taylor Whittaker and George Neville Watson

[WW 69] *A Course of Modern Analysis*, fourth edition, Cambridge University Press, 1969, p. 133.

William Henry Young

[Yo 09a] *The Fundamental Theorems of the Differential Calculus*, Hafner Publishing Company, New York, 1909.

[Yo 09b] On implicit functions and their differentials, *Proceedings of the London Mathematical Society* 7 (1909), 397–421.

Eberhard Zeidler

[Ze 86] *Nonlinear Functional Analysis and its Applications*, volume 1, Springer-Verlag, New York, 1986.

Index